# Lecture Notes in Bioinformatics     **8452**

Subseries of Lecture Notes in Computer Science

More information about this series at http://www.springer.com/series/5381

Enrico Formenti · Roberto Tagliaferri
Ernst Wit (Eds.)

# Computational Intelligence Methods for Bioinformatics and Biostatistics

10th International Meeting, CIBB 2013
Nice, France, June 20–22, 2013
Revised Selected Papers

 Springer

*Editors*
Enrico Formenti
University Nice Sophia Antipolis
Sophia Antipolis
France

Ernst Wit
University of Groningen
AG Groningen
The Netherlands

Roberto Tagliaferri
University of Salerno
Fisciano
Italy

ISSN 0302-9743          ISSN 1611-3349  (electronic)
ISBN 978-3-319-09041-2  ISBN 978-3-319-09042-9  (eBook)
DOI 10.1007/978-3-319-09042-9

Library of Congress Control Number: 2014945214

LNCS Sublibrary: SL8 – Bioinformatics

Springer Cham Heidelberg New York Dordrecht London

Printed on acid-free paper

Springer is part of Springer Science+Business Media (www.springer.com)

# Preface

This volume contains the proceedings of the International Meeting on Computational Intelligence Methods for Bioinformatics and Biostatistics (CIBB 2013), which was in its tenth edition this year. While many past editions were organized in Italy, from last year when the conference took place at the Methodist Research Institute, Houston (USA), the conference started an internationalization process. It coincided with a further enlargement of the spectrum of the scientific domains covered.

Indeed, the main scope of the CIBB meeting series is to provide a forum open to researchers from different disciplines to present and discuss problems concerning computational techniques in bioinformatics, systems biology, and medical and health informatics with a particular focus on neural networks, machine learning, fuzzy logic, and evolutionary computation methods.

This year CIBB was co-located and followed the PRIB (Pattern Recognition In Bioinformatics) conference and there were many authors who contributed to both conferences. As organizers of CIBB we would like to thank all the contributors and organizers of PRIB. There was also a common day between the two conferences. We hope that this was an occasion for new scientific collaborations and exchanges. Many thanks also go to the invited speakers: Sylvain Sené (Aix-Marseille Université, France) Anne Siegel (IRISA CNRS and Inria Rennes, France) and Ernst Wit (University of Groningen, The Netherlands) for their excellent talks.

This year 33 papers were selected for presentation at the conference, and each paper received two reports on average. A further reviewing process took place for the 19 papers that were selected to appear in this volume. The authors are spread over more than ten different countries: Algeria (3), Canada (2), France (18), Islamic Republic of Iran (4), Italy (55), The Netherlands (1), Norway (1), Romania (6), Taiwan (2), Tunisia (2), UK (6), and USA (7).

The editors would like to thank all the Program Committee members and the external reviewers both of the conference and post-conference version of the papers for their valuable work. We are also indebted to the chairs of the very interesting and successful special sessions ("Knowledge-Based Medicine" and "Data Integration and Analysis in Omic-Science"), which attracted even more contributions and attention.

A big thanks also to the munificent sponsors, and in particular to Nice Sophia Antipolis University, which made this event possible. And last but not least, the editors would also like to thank all the authors for the high quality of the papers they contributed and warmly invite them to submit their work to the next edition that will take place in Cambridge.

March 2013

Enrico Formenti
Roberto Tagliaferri
Ernst Wit

# Organization

## General Chairs

Enrico Formenti    Nice Sophia Antipolis University, France
Roberto Tagliaferri    University of Salerno, Italy
Ernst Wit    University of Groningen, The Netherlands

## Special Sessions Chairs

Claudia Angelini    IAC-CNR, Italy
Elia Biganzoli    Università degli Studi di Milano, Italy
Clelia Di Serio    Università Vita-Salute San Raffaele, Italy
Alexandru Floares    Oncological Institute Cluj-Napoca, Romania
Leif Peterson    Houston Methodist Research Institute, USA
Alfredo Vellido    Universitat Politècnica de Catalunya, Spain

## Program Committee

Federico Ambrogi    University of Milan, Italy
Sansanee Auephanwiriyakul    Chiang Mai University, Thailand
Sanghamitra Bandyopadhyay    Indian Statistical Institute, Kolkata, India
Gilles Bernot    Nice Sophia Antipolis University, France
Chengpeng Bi    Childrens Mercy Hospital, Kansas City, USA
Mario Cannataro    University of Magna Graecia, Catanzaro, Italy
Virginio Cantoni    Università di Pavia, Italy
Xue-Wen Chen    University of Kansas, Lawrence, USA
Adele Cutler    Utah State University, Logan, USA
Paolo Decuzzi    TMHRI, Houston, Texas, USA
Angelo Facchiano    Istituto di Scienze dell'Alimentazione - CNR, Italy
Leonardo Franco    University of Malaga, Spain
Christoph Friedrich    University of Applied Science and Arts, Dortmund, Germany
Raffaele Giancarlo    University of Palermo, Italy
Saman K. Halgamuge    The University of Melbourne, Australia
Emmanuel Ifeachor    University of Plymouth, UK
Mika Sato-Ilic    University of Tsukuba, Japan
Paulo Lisboa    Liverpool John Moores University, UK
Vincenzo Manca    Università di Verona, Italy
Elena Marchiori    Radboud University, Nijmegen, The Netherlands
Giancarlo Mauri    Università degli Studi di Milano-Bicocca, Italy

| Luciano Milanesi | ITB-CNR, Milan, Italy |
| Taishin Nomura | Osaka University, Osaka, Japan |
| Carlos-Andres Pena-Reyes | University of Applied Sciences Western Switzerland, Switzerland |
| Vassilis Plagianakos | University of Central Greece, Lamia, Greece |
| Riccardo Rizzo | ICAR-CNR, Palermo, Italy |
| Paolo Romano | National Cancer Research Institute, Genoa, Italy |
| Stefano Rovetta | University of Genova, Italy |
| Jianhua Ruan | University of Texas, San Antonio, USA |
| Luis Rueda | University of Windsor, Canada |
| Andrey Rzhetsky | University of Chicago, USA |
| Jennifer Smith | Boise State University, USA |
| Giorgio Valentini | University of Milan, Italy |
| Alfredo Vellido | Universidad Politecnica de Catalunya, Spain |
| Yanqing Zhang | Georgia State University, Atlanta, USA |

## Steering Committee

| Thomas Back | Leiden University, The Netherlands |
| Pierre Baldi | University of California, Irvine, USA |
| Elia Biganzoli | University of Milan, Italy |
| Alexandru Floares | Oncological Institute Cluj-Napoca, Romania |
| Jon Garibaldi | University of Nottingham, UK |
| Nikola Kasabov | Auckland University of Technology, New Zealand |
| Francesco Masulli | University of Genova, Italy and Temple University, USA |
| Leif Peterson | TMHRI, Houston, Texas, USA |
| Roberto Tagliaferri | University of Salerno, Italy |

## Organizing Committee

| Enrico Formenti (Chair) | Nice Sophia Antipolis University, France |
| Sandrine Julia | Nice Sophia Antipolis University, France |
| Corinne Jullien | Nice Sophia Antipolis University, France |
| Bruno Martin | Nice Sophia Antipolis University, France |
| Christophe Papazian | Nice Sophia Antipolis University, France |
| Julien Provillard | Nice Sophia Antipolis University, France |
| Magali Richir | Nice Sophia Antipolis University, France |

## External Reviewers

| Claudia Angelini | Daniela De Canditiis |
| Gilles Bernot | Angelo Facchiano |
| Elia Biganzoli | Mario Guarracino |

Michele La Rocca
Luca Manzoni
Francesco Masulli
Marco Muselli
Mathilde Noual

Nicolas Pasquier
Giancarlo Raiconi
Riccardo Rizzo
Ernst Wit

## Sponsors

We thank very deeply the sponsors that made CIBB 2013 possible (alphabetical order):

Società Italiana di Bioinformatica

Centre National de la Recherche Scientifique

Dipartimento di Informatica, Università di Salerno, Italy

Ecole Doctorale STIC

Projet ANR EMC (ANR-09-BLAN-0164)

Laboratoire d'Informatique, Signaux et Systèmes de Sophia Antipolis, France

INNS-SIG on Bioinformatics and Intelligence

Italian Network for Oncology Bioinformatics
Rete Nazionale di Bioinformatica Oncologica

Italian Network for Oncology Bioinformatics

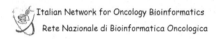

Società Italiana REti Neuroniche

# Contents

## Special Session: Knowledge Based Medicine

## Special Session: Data Integration and Analysis in Omic-Science

# Keynote Speaker

# Dynamic Gaussian Graphical Models
# for Modelling Genomic Networks

Antonio Abbruzzo[1], Clelia Di Serio[2], and Ernst Wit[3]($\boxtimes$)

[1] University of Palermo, Palermo, Italy
A.Abbruzzo@unipa.it
[2] Universita Vita-Salute San Raffaele, Milano, Italy
[3] Johann Bernoulli Institute, University of Groningen, Groningen, The Netherlands
e.c.wit@rug.nl

**Abstract.** After sequencing the entire DNA for various organisms, the challenge has become understanding the functional interrelatedness of the genome. Only by understanding the pathways for various complex diseases can we begin to make sense of any type of treatment. Unfortunately, deciphering the genomic network structure is an enormous task. Even with a small number of genes the number of possible networks is very large. This problem becomes even more difficult, when we consider dynamical networks. We consider the problem of estimating a sparse dynamic Gaussian graphical model with $L_1$ penalized maximum likelihood of structured precision matrix. The structure can consist of specific time dynamics, known presence or absence of links in the graphical model or equality constraints on the parameters. The model is defined on the basis of partial correlations, which results in a specific class precision matrices. A priori $L_1$ penalized maximum likelihood estimation in this class is extremely difficult, because of the above mentioned constraints, the computational complexity of the $L_1$ constraint on the side of the usual positive-definite constraint. The implementation is non-trivial, but we show that the computation can be done effectively by taking advantage of an efficient maximum determinant algorithm developed in convex optimization.

## 1 Introduction

Networks are important models to address specific questions in genomics. Dynamic gene-regulatory networks are complex objects since the number of potential components involved in the system is very large. For example, one important direction in systems biology is to discover gene regulatory networks from microarray data based on the observed mRNA levels of thousands of genes under various conditions. We shall show that one solution to such problem is the use of penalized Gaussian graphical models, which have been extensively used to estimate sparse static graphs.

*Proteins* are essential parts of the cell that determine the cell's structure and execute nearly all its functions. The production of proteins is carried out by the

© Springer International Publishing Switzerland 2014
E. Formenti et al. (Eds.): CIBB 2013, LNBI 8452, pp. 3–12, 2014.
DOI: 10.1007/978-3-319-09042-9_1

*ribosomes*, but the information needed for their production is encoded in *genes* which are the segments of *DNA*. DNA contains valuable genetic information, that must be preserved. Transient *RNA* is used to carry the message from DNA to ribosomes. In all living cells, the flow of genetic information is thought to go in this way

$$DNA \rightarrow RNA \rightarrow \text{PROTEIN}.$$

This fundamental principle in biology is called the *central dogma* of molecular biology. The step from DNA to RNA consists of copying the information from genes to RNA and it is called *transcription*. The step from RNA to protein consists of decoding the information from RNA by ribosomes and it is called *translation*. Together these two processes are known as *gene expression*.

The process of transcription is carried out by special enzymes called *RNA polymerases* (RNAp). RNA polymerase binds to the promoter and then opens up the double helix of the DNA sequence immediately in front of it and slides down the gene producing the RNA molecule. The *promoter* is a region of DNA that facilitates the transaction of a particular gene and contains a sequence of nucleotides indicating the starting point for RNA synthesis. Chain elongation continues until enzyme encounters a second signal in DNA, the *terminator*, where RNAp halts and releases both the DNA chain and the newly made RNA chain. RNA which encodes information for production of a certain protein is called *messenger RNA*(mRNA).

However, to do all of this RNAp needs help from special proteins called *transcription factors*. Transcription factors bind at the promoter and form a transcription initiation complex. They position the RNAp correctly on the promoter and aid in pulling apart the two strands of DNA to allow transcription to begin and to allow RNAp to leave promoter as transcription begins. After RNAp is released from the complex it starts making RNA. Once transcription has begun, most of the transcription factors are released from the DNA so that they are available to initiate another round of transcription with a new RNAp molecule. The synthesis of the next RNA usually starts before the first RNA is completed. There maybe several polymerases moving along a single stretch of DNA and RNAs.

The main goal of gene transcription is to produce mRNA which will be translated by ribosomes to make proteins. Each mRNA can be translated several times by ribosome in order to make proteins. This is done until mRNA reaches the end of its life-span. The network of gene regulation can be very complex, where one regulatory protein controls genes that produce other regulators that in turn control other genes. Gene regulatory network models can be represented as directed or undirected graphs, where nodes are the elements, such as DNA, RNA, proteins etc. The directed or undirected edges from one node to another represent the corresponding interaction, for example, activation, repression or translation. Being able to create gene regulatory networks from experimental data and to use them to think about their dynamics is the aim of this paper.

## 2   Graphical Models

An *undirected graphical model* is also called a Markov random field. It is defined as a pair $(G, \mathbb{P})$ that specifies a probability density function $f$ for their joint distribution $\mathbb{P}$ in the form

$$(F) \qquad f(y_1, \ldots, y_p) = \frac{1}{z} \prod_{c \in C} \psi_c(\mathbf{y}_c), \qquad (1)$$

where $C$ is a set of cliques, i.e. complete subsets of $V$ that are maximal, in $G$, $\psi_c(\mathbf{y}_c)$ is a potential function, which is a positive function of the variables $\{y_i\}_{i \in C}$, and

$$z = \sum_{\mathbf{y}} \prod_{c \in C} \psi_c(\mathbf{y}_c)$$

is a normalization factor. If the factorization (F) is possible, then it implies the global Markov property. A probability distribution $\mathbb{P}$ is said to obey the *global Markov property*, relative to $G$, if for any triple $(A,B,S)$ of disjoint subsets of $V$ such that $S$ separates $A$ from $B$ in $G$

$$(G) \qquad\qquad \mathbf{Y}_A \perp \mathbf{Y}_B | \mathbf{Y}_S.$$

The global Markov property in turn implies the local and pairwise Markov properties. A probability distribution function is said to obey:

(L) the *local Markov property*, relative to $G$, if for any vertex $i \in V$

$$Y_i \perp \mathbf{Y}_{V \setminus \{cl(i)\}} | \mathbf{Y}_{bd(i)},$$

(P) the *pairwise Markov property*, relative to $G$, if for any pair $(i,j)$ of non-adjacent vertices

$$Y_i \perp Y_j | \mathbf{Y}_{V \setminus \{i,j\}},$$

The boundary of $i$ is the set of nodes such that $bd(i) = pa(i) \cup ne(i)$, and the closure of $i$ is the set of nodes such that $cl(i) = i \cup bd(i)$. The expression $V \setminus \{i,j\}$ indicates the set of nodes $V$ except nodes $i$ and $j$. The expression $Y_i \perp Y_j | \mathbf{Y}_{V \setminus \{i,j\}}$ means that the probability distribution function can be factorized as follows:

$$f_{Y_i, Y_j | \mathbf{Y}_{V \setminus \{i,j\}}}(y_i, y_j | \mathbf{y}_{V \setminus \{i,j\}}) = f_{Y_i | \mathbf{Y}_{V \setminus \{i,j\}}}(y_i | \mathbf{y}_{V \setminus \{i,j\}}) f_{Y_j | \mathbf{y}_{V \setminus \{i,j\}}}(y_j | \mathbf{y}_{V \setminus \{i,j\}}).$$

It can be shown that $(F) \Rightarrow (G) \Rightarrow (L) \Rightarrow (P)$ [1]. Moreover, Hammersley and Clifford's theorem states that:

**Theorem 1 (Hammersley and Clifford).** *A probability distribution $\mathbb{P}$ with positive and continuous density $f$ with respect to a product measure $\mu$ satisfies the pairwise Markov property with respect to an undirected graph $G$ if and only if it factorizes according to $G$.*

This theorem gives the necessary and sufficient condition for $(P) \Leftrightarrow (F)$, and under this condition we have that all Markov properties are equivalent:

$$(F) \Leftrightarrow (G) \Leftrightarrow (L) \Leftrightarrow (P).$$

Undirected graphical models are useful when random variables can be analysed symmetrically. Specific undirected graphical models are distinguished by the choice of the undirected graph $G$ and the potential functions $\psi_c$.

A multivariate Gaussian graphical model (GGM) for an undirected graph $G$ is defined in terms of its Markov properties. Variables, i.e. nodes in the graph, are independent conditional on a separating set. In other words, let $X = (X_1, X_2, \ldots, X_p)^T$ be a multivariate Gaussian vector, then an undirected edge is drawn between two nodes $i$ and $j$, if and only if the corresponding variables $X_i$ and $X_j$ are conditionally dependent given the remaining variables. Let $G = (X, E)$ be an undirected graph with vertex set $X = \{X_1, \ldots, X_p\}$ and edge set $E = \{e_{ij}\}$, where $e_{ij} = 1$ or $0$ according to whether vertices $i$ and $j$ are adjacent in $G$ or not. The GGM model $N(G)$ consists of all p-variate normal distributions $N_p(\mu, \Sigma)$, for arbitrary mean vectors $\mu$ and covariance matrices $\Sigma$, assumed nonsingular, for which the concentration or precision matrix $\Theta = \Sigma^{-1}$ satisfies the linear restriction $e_{ij} = 0 \Leftrightarrow \theta_{ij} = 0$.

The model $N(G)$ has also been called a covariance selection model [2] and a concentration graph model [3]. The reader is referred to [4, Chap. 6] for statistical properties of these models, including methods for parameter estimation, model testing and model selection. The model $N(G)$ also can be defined in terms of pairwise conditional independence. If $X = (X_1, \ldots, X_p)^T \sim N_p(\mu, \Sigma)$, then

$$\theta_{ij} = 0 \Leftrightarrow X_j \perp X_i | X_{\{-(i,j)\}} \Leftrightarrow \rho_{ij} = 0$$

where $\rho_{ij} = -\theta_{ij}/\sqrt{\theta_{ij}\theta_{ij}}$ denotes the partial correlation between $X_i$ and $X_j$, i.e. the correlation between $X_i$ and $X_j$ given $X_{\{-(i,j)\}}$. This suggests that the determination of the graph $G$, can be based on the set of sample partial correlations $\hat{\rho}_{ij}$ arising from independent and identically distributed observations $X \sim N_p(\mu, \Sigma)$, where $n >> p$ is assumed in order to guarantee positive definiteness of the sample covariance matrix. In other words, given a random sample $X$ we wish to estimate the concentration matrix $\Theta$. Of particular interest is the identification of zero entries in the concentration matrix $\Theta = \{\theta_{ij}\}$, since a zero entry $\theta_{ij} = 0$ indicates the conditional independence between the two variables $X_i$ and $X_j$ given all other variables.

Graphical models are probability models for multivariate random variables whose independence structure is characterized by a conditional independence graph. The standard theory of estimating GGMs can be exploited only when the number of measurements $n$ is much higher than the number of variables $p$. This ensures that the sample covariance matrix is positive definite with probability one. Instead, in most application, such as microarray gene expression data sets, we have to cope with the opposite situation ($n \ll p$). Thus, the growing interest in "small $n$, large $p$" problems, requires an alternative approach. In problems where the number of nodes is large, but the number of links are relatively few per node, sparse inference of $\Theta$ in the framework of a GGM is useful.

Estimating the dimensionality of the GGM model is complicated issue. The standard approach is greedy stepwise forward-selection or backward-deletion, and parameter estimation is based on the selected model. In each step the edge selection or deletion is typically done through hypothesis testing at some level $\alpha$. It has long been recognized that this procedure does not correctly take account of the multiple comparisons involved [5]. Another drawback of the common stepwise procedure is its computational complexity. To remedy these problems, [6] proposed a method that produces conservative simultaneous $1 - \alpha$ confidence intervals, and use these confidence intervals to do model selection in a single step. The method is based on asymptotic considerations. Reference [7] proposed a computationally attractive method for covariance selection that can be used for very large Gaussian graphs. They perform neighbourhood selection for each node in the graph and combine the results to learn the structure of a Gaussian concentration graph model. They showed that their method is consistent for sparse high-dimensional graphs. However, in all of the above mentioned methods, model selection and parameter estimation are done separately. The parameters in the concentration matrix are typically estimated based on the model selected. As demonstrated by [8], the discrete nature of such procedures often leads to instability of the estimator: small changes in the data may result in very different estimates.

Here, we propose a sparse dynamic Gaussian graphical model with $L_1$ penalty of structured correlation matrix that does model selection and parameter estimation simultaneously in the Gaussian concentration graph model. We employ an $L_1$ penalty on the off-diagonal elements of the correlation matrix. This is similar to the idea of the *glasso* [9]. The $L_1$ penalty encourages sparsity and at the same time gives shrinkage estimates. In addition, we can model arbitrary, locally additive models for the precision matrix, while explicitly ensuring that the estimator of the concentration matrix is positive define.

## 3   Dynamic Gaussian Graphical Model for Networks

The graph structure of the Gaussian graphical model describes the conditional independence structure between the variables. The two main applications of this conditional independence are either (i) modular dependency structures and (ii) Markovian dependency structures. The former are used in expert systems or flowchart descriptions of causal structures, whereas the latter is typical for spatio-temporal forms of (in)dependence. A dynamic gaussian graphical model for a network contains both types of conditional dependence: a Markovian dependence structure would capture that temporal relatedness of nearby observations, which is broken by one (or more) conditioning, intervening observations. The network itself has an internal relatedness due to the modular structure of the network: the results of the observed outcomes at the nodes flow through the links to the other nodes, thereby affecting neighbouring vertices. Due to its computational tractability is the multivariate normal distribution uniquely suited as an initial model for a dynamic graphical model. If we measure a univariate outcome at $p$ nodes across $T$ discrete time-points, then initially we describe the data $X$ as coming from a multivariate normal distribution:

$$X \sim N_{pT}(\mu, \Theta^{-1}).$$

In many practical example, it may be the case that only a single replicate $X$ has been observed. Estimation will only be possible if we are willing to impose restrictions on the parameters. There are two types of restrictions that we will consider: sparsity restrictions and model definitions.

### 3.1   Sparsity Restrictions of the Precision Matrix

The arrival of the high-throughput era in genomics has seen an explosion of data gathering: for a fraction of the amount of time and money it used to cost to monitor the level of a particular gene or protein, now thousands are monitored. Nevertheless, the underlying physical reality will not have changed as a result of our data-gathering. The particular protein that used to bind to the promotor region of the particular gene will still do so: the fact that we monitor thousands of genomic variables has not made the genomic reality itself any more difficult. Obviously, this reality is certainly highly complex, but at the same time it is also highly structured as DNA sequences are highly specific for binding to particular proteins. Therefore, the genomic network can be thought to be highly sparse set of relations between thousands of genomic players, such as DNA, mRNA and proteins. Obviously, we don't know exactly which links should be assumed to be zero, but we want to create a model that encourages zeroes between the vertices.

Furthermore, the fact that we are considering dynamic models with observations of the genomic system spaced in time, it is probably sufficient to assume – especially given the usual spacing of genomic observations – the existence of first or at most second order Markov dependence. This means that large part of the precision matrix can be filled with zeroes *a priori*.

### 3.2   Model Restrictions of the Precision Matrix

Given the sparsity of the data, it is essential to define models that are finely tuned to be able to estimate interesting quantities of interest. For example, we have seen in the previous paragraph that Markov assumptions are sensible ways to reduce the dimensionality of the estimation problem. Additionally, given that the temporal correlation is probably not particularly important, it makes sense to compromise a little on the amount of variables we use to model it. For example, it makes sense to restrict the attention to models in which

$$\forall i, t : \mathrm{cor}(x_{i,t}, x_{i,t-1} | x_{-i}) = \rho.$$

This reduces the number of parameters in $\Theta$ by $pT - 1$. Moreover, it may, in certain circumstances, be sensible to assume that the genomic network at each time-point is the same. This reduces the number of parameters by $(T - 1)p^2$.

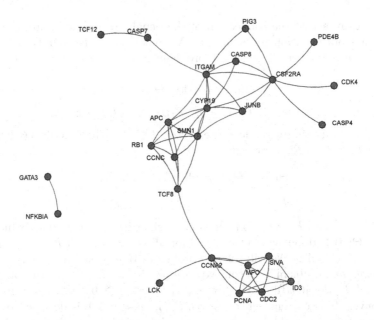

**Fig. 1.** The lag zero network selected in the case of the T-cell data. It shows two hubs involving the JUNB and CCNC genes, which are well-known for being central regulator. Blue and red links represent positive and negative partial correlations, respectively (color figure online).

### 3.3    Maximum Likelihood

The most simple model is the unconstrained $\Theta$ with no penalty on the elements $\theta_{ij}$ on the precision matrix $\Theta$. The log-likelihood for $\mu$ and $\Theta = \Sigma^{-1}$ based on a random sample $X = (X^{(1)}, \ldots, X^{(n)})$ is $l(\mu, \Sigma; X) \cong \frac{n}{2} \log |\Theta| - \frac{1}{2} \sum_{i=1}^{n} (X_i - \mu)^T \Theta (X_i - \mu)$ up to a constant not depending on $\mu$ and $\Theta$. Even if $S = \frac{1}{n} \sum_{i=1}^{n} (X_i - \bar{X})(X_i - \bar{X})^T$ is of full rank (only if $n > pT$), the matrix $S^{-1}$ will not be 'sparse'. To achieve 'sparse' graph structure and to obtain a better estimator of the concentration matrix, we introduce an $L_1$ penalty on the likelihood, i.e. we want a minimizer $\Theta$ of

$$- \log |\Theta| + \text{trace} \, (\Sigma S) \text{ subject to} \sum_{i \neq j} |\theta_{ij}| \leq t, \qquad (2)$$

over the set of positive definite matrices $\Theta$. Here $t \geq 0$ is a tuning parameter.

The constraint as formulated above does not penalize the diagonal of $\Theta$. We could also choose not to penalize links that we know are there or time-dependencies which are so low-dimensional that it is not worth penalizing.

## 4    Max Determinant Optimization Problem

The non-linearity of the objective function, the positive definiteness constraint and the structured correlation make the optimization problem non-trivial.

We take advantage of the connection of the penalized likelihood and the max-determinant optimization problem [10]. We make use of the SDPT3 algorithm [11] to manage higher dimensional problems. We consider the optimization problem:

$$\min \quad c^T \beta + \log |\Theta(\beta)| \tag{3}$$

$$\text{subject to } \Theta(\beta) \geq 0, \quad F(\beta) \geq 0, \quad L\beta = b;$$

where the optimization variable is the vector $\beta \in R^m$. The functions $\Theta : R^m \rightarrow R^{l \times l}$ and $F : R^m \rightarrow R^{n \times n}$ are affine:

$$\Theta(\beta) = \Theta_0 + \beta_1 \Theta_1 + \ldots + \beta_m \Theta_m$$

$$F(\beta) = F_0 + \beta_1 F_1 + \ldots + \beta_m F_m,$$

where $\Theta_i = \Theta^T$ and $F_i = F_i^T$. The inequality signs in (3) denote matrix inequalities, i.e., $\Theta(\beta) > 0$ means $z^T \Theta(\beta) z \geq 0$ for all nonzero $z$ and $F(\beta) \geq 0$ means $z^T F(\beta) z \geq 0$ for all $z$. We will refer to problem (3) as a *maxdet* problem.

The *maxdet* problem is a convex optimization problem, i.e. the objective function $c^T \beta + \log |\Theta(\beta)|$, is convex (*on* $\{x : \Theta(\beta) \geq 0\}$, and the constraint set is convex. The current version of *SDPT3*, version 4.0, is designed to solve conic programming problems whose constraint cone is a product of semidefinite cones, second-order cones, nonnegative orthants and Euclidean spaces; and whose objective function is the sum of linear functions and log-barrier terms associated with the constraint cones.

## 5   Application to T-Cell Data

Tcell dataset is a large time-series experiment to characterize the response of a human T-cell line (Jurkat) to PMA and ionomycin treatment. The data set contains the temporal expression levels of 57 genes for 10 unequally spaced time points. At each time point there are 44 separate measurements. See [12] for more details.

We consider a particular structure to the graphical model. We define the nodes of the graph to be the genes at a particular time point. This results in a $570 \times 570$ inverse covariance matrix $\Theta$. This requires estimating more than 160,000 parameters with only 25,000 observations. However, there is good reason to impose some constraints on $\Theta$.

1. **Markov assumption:** we assume that except for lag zero and lag one, there are no higher order interactions between the genes, i.e.,

$$Cov(X_{gt}, X_{g't'}) = 0 \text{ for } |t - t'| > 1.$$

2. **Interaction persistence:** For the lag zero and lag one interactions, we assume that the interactions are persistent across all ten time points, i.e.,

$$\text{Lag 0: } \Omega_{gt,g't} = \Omega_{gs,g's},$$

$$\text{Lag 1: } \Omega_{gt,g't+1} = \Omega_{gs,g's+1}.$$

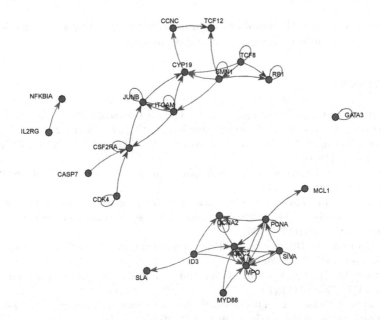

**Fig. 2.** The lag one network for the T-cell data: the arrows are a semantic interpretation of the graphical model. They are given their direction by pointing from the past to the future, although in the structure of the graphical model they are in fact undirected. Blue and red links represent positive and negative partial correlations, respectively (color figure online).

This reduces the number of parameters from over 160,000 to a manageable number less than 5,000. Furthermore, the shrinkage induced by the $L_1$ penalty further stabilizes the estimates. The application of the above model to the T-cell data, results in the lag zero graph shown in Fig. 1 and the lag one graph shown in Fig. 2. Blue and red links represent positive and negative partial correlations, respectively. We see a typicall feature that the majority of links are blue, as it is impossible to have stable networks with *a lot* of negative interactions. Furthermore, the networks we infer seem to have other typical characteristics of genomic networks, such as modularity and small world properties.

## 6    Conclusions

As more and more large datasets become available, the need for efficient tools to analyse such data has become imperative. In this paper, we have considered sparse dynamic Gaussian graphical models with $\ell_1$-norm penalty. This type of modelling offers a straightforward interpretation: the edges of the graph define the partial conditional correlations among the nodes. In particular, under the sparsity assumption, a large part of the precision matrix can be filled with zeroes a priori. Based on the consideration of dynamic and model-oriented definitions, we are able to reduce the number of parameters to be estimated, which allows for more relevant interpretations in real data analysis.

# References

1. Lauritzen, S.: Graphical Models, vol. 17. Oxford University Press, USA (1996)
2. Dempster, A.: Covariance selection. Biometrics **28**, 157–175 (1972)
3. Cox, D., Wermuth, N.: Multivariate Dependencies. Chapman and Hall/CRC Press, Boca Raton (1996)
4. Whittaker, J.: Graphical Models in Applied Multivariate Statistics, vol. 16. Wiley, New York (1990)
5. Edwards, D.: Introduction to Graphical Modelling. Springer, New York (2000)
6. Drton, M., Perlman, M.: Model selection for Gaussian concentration graphs. Biometrika **91**(3), 591–602 (2004)
7. Meinshausen, N., Bühlmann, P.: High-dimensional graphs with the Lasso. Ann. Statist. **34**, 1436–1462 (2006)
8. Breiman, L.: Heuristics of instability and stabilization in model selection. Ann. Stat. **24**(6), 2350–2383 (1996)
9. Friedman, J., Hastie, T., Tibshirani, R.: Sparse inverse covariance estimation with the graphical lasso. Biostatistics **9**(3), 432 (2007)
10. Boyd, S., Vanderberghe, L.: Convex optimization. Cambridge Univ. Pr., New York (2004)
11. Toh, K.-C., Todd, M.J., Tütüncü, R.H.: On the implementation and usage of *SDPT*3 - a MATLAB software package for semidefinite quadratic linear programming, version 4.0. In: Anjos, M.F., Lasserre, J.B. (eds.) Handbook on Semidefinite, Conic and Polynomial Optimization. International Series in Operations Research and Management Science, vol. 166, pp. 715–754. Springer, NewYork (2006)
12. Rangel, C., Angus, J., Ghahramani, Z., Lioumi, M., Sotheran, E., Gaiba, A., Wild, D., Falciani, F.: Modeling T-cell activation using gene expression profiling and state-space models. Bioinformatics **20**(9), 1361–1372 (2004)

# Bioinformatics Regular Session

# Molecular Docking for Drug Discovery: Machine-Learning Approaches for Native Pose Prediction of Protein-Ligand Complexes

Hossam M. Ashtawy and Nihar R. Mahapatra[✉]

Department of Electrical and Computer Engineering,
Michigan State University, East Lansing, MI 48824, USA
{ashtawy,nrm}@egr.msu.edu

**Abstract.** Molecular docking is a widely-employed method in structure-based drug design. An essential component of molecular docking programs is a scoring function (SF) that can be used to identify the most stable binding pose of a ligand, when bound to a receptor protein, from among a large set of candidate poses. Despite intense efforts in developing conventional SFs, which are either force-field based, knowledge-based, or empirical, their limited docking power (or ability to successfully identify the correct pose) has been a major impediment to cost-effective drug discovery. Therefore, in this work, we explore a range of novel SFs employing different machine-learning (ML) approaches in conjunction with physicochemical and geometrical features characterizing protein-ligand complexes to predict the native or near-native pose of a ligand docked to a receptor protein's binding site. We assess the docking accuracies of these new ML SFs as well as those of conventional SFs in the context of the 2007 PDBbind benchmark datasets on both diverse and homogeneous (protein-family-specific) test sets. We find that the best performing ML SF has a success rate of 80 % in identifying poses that are within 1 Å root-mean-square deviation from the native poses of 65 different protein families. This is in comparison to a success rate of only 70 % achieved by the best conventional SF, ASP, employed in the commercial docking software GOLD. We also observed steady gains in the performance of the proposed ML SFs as the training set size was increased by considering more protein-ligand complexes and/or more computationally-generated poses for each complex.

## 1 Introduction

### 1.1 Background

Bringing a new drug to the market is a complex process that costs hundreds of millions of dollars and spans over ten years of research, development, and testing. A fairly big portion of this hefty budget and long time-line is spent in the early stages of drug design that involves two main steps: first, the enzyme, receptor, or other protein responsible for a disease of interest is identified; second, a small

© Springer International Publishing Switzerland 2014
E. Formenti et al. (Eds.): CIBB 2013, LNBI 8452, pp. 15–32, 2014.
DOI: 10.1007/978-3-319-09042-9_2

molecule or *ligand* is found or designed that will bind to the target protein, modulate its behavior, and provide therapeutic benefit to the patient. Typically, *high-throughput screening* (HTS) facilities with automated devices and robots are used to synthesize and screen ligands against a target protein. However, due to the large number of ligands that need to be screened, HTS is not fast and cost-effective enough as a lead identification method in the initial phases of drug discovery [1]. Therefore, computational methods referred to as *virtual screening* are employed to complement HTS by narrowing down the number of ligands to be physically screened. In virtual screening, information such as structure and physicochemical properties of a ligand, protein, or both, are used to estimate both *binding pose* and/or *binding affinity*, which represents the strength of association between the ligand and its receptor protein. The most popular approach to predicting the correct binding pose and binding affinity (BA) in virtual screening is *structure-based* in which physicochemical interactions between a ligand and receptor are deduced from the 3D structures of both molecules. This *in silico* method is also known as *protein-based* as opposed to the alternative approach, *ligand-based*, in which only ligands that are biochemically similar to the ones known to bind to the target are screened.

In this work, our focus will be on protein-based drug design, wherein ligands are placed into the active site of the receptor. The 3D structure of a ligand, when bound to a protein, is known as *ligand active conformation. Binding mode* refers to the orientation of a ligand relative to the target and the protein-ligand conformation in the bound state. A binding pose is simply a candidate binding mode. In molecular *docking*, a large number of binding poses are computationally generated and then evaluated using a *scoring function (SF)*, which is a mathematical or predictive model that produces a score representing binding stability of the pose. The outcome of the docking run, therefore, is a ligand's top pose ranked according to its predicted binding score as shown in Fig. 1. Typically, this docking and scoring step is performed iteratively over a database containing thousands to millions of ligand candidates. After predicting their binding poses, another scoring round is performed to rank ligands according to their predicted binding free energies. The top-ranked ligand, considered the most promising drug candidate, is synthesized and physically screened using HTS.

The most important steps in the docking process are scoring ligands' conformations at their respective binding sites and ranking ligands against each other. These core steps affect the outcome of the entire drug search campaign. That is because predictions of scoring functions determine which binding orientation/conformation is deemed the best, which ligand from a database is considered likely to be the most effective drug, and the estimated binding affinity (BA). Correspondingly, three main capabilities that a reliable scoring function should have are: (i) the ability to identify the correct binding mode of a ligand from among a set of (computationally-generated) poses, (ii) the ability to correctly rank a given set of ligands, with known binding modes when bound to the same protein, and, finally, (iii) the ability to produce binding scores that are (linearly) correlated to the experimentally-determined binding affinities of

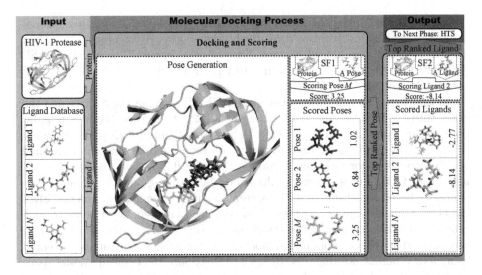

**Fig. 1.** Protein-ligand docking and ranking workflow.

protein-ligand complexes with known 3D structures. These three performance attributes were referred to by Cheng et al. as *docking power*, *ranking power*, and *scoring power*, respectively [2]. We refer to the corresponding problems as *ligand docking*, *ligand ranking*, and *ligand scoring problems*. In practice and in all existing work, a single general SF is trained to predict protein-ligand BA and then used in both the ligand docking and ranking stages to identify the top pose and ligand, respectively. In this work, we propose docking-specialized machine-learning SFs capable of predicting native poses more accurately than the conventional BA-based SFs. These native-pose prediction models are used as SF1 in Fig. 1. As for the second scoring round, designated by SF2 in Fig. 1, in previous work we built accurate machine-learning SFs to score and rank ligands against each other using their predicted binding affinities [3,4].

## 1.2   Related Work

Most SFs in use today can be categorized as either *force-field-based* [5], *empirical* [6], or *knowledge-based* [7] SFs. Despite intense efforts into these conventional scoring schemes, several recent studies report that the docking power of existing SFs is quite limited. Cheng and co-workers recently conducted an extensive test of sixteen SFs from these three categories that are either employed in mainstream commercial docking tools and/or have been developed in academia [2]. The main test set used in their study consisted of 195 diverse protein-ligand complexes and four other protein-specific test sets. In order to assess the docking power of all SFs, they generated 100-pose decoy sets for each protein-ligand complex in the main test set. They defined the *docking power* of an SF as its rate of success in identifying binding poses that are within a certain root-mean-square deviation (RMSD) from the native pose over all complexes. Using this criteria, three SFs were found to have a relatively higher level of accuracy when their docking abilities were judged in three different experiments. These SFs are ASP [8] in the GOLD [9] docking software, PLP1 [10] in Discovery Studio [11], and the stand-alone SF DrugScore [12].

In this work, we will compare our novel ML SFs against these three and the other thirteen SFs considered by Cheng et al. [2]. They used the four popular docking programs LigandFit [13], GOLD, Surflex [14], and FlexX [15] to generate diverse sets of decoy poses. Each of these tools employs different conformational search algorithms for best poses. Namely, LigandFit relies on a shape-directed algorithm, GOLD uses a genetic algorithm, Surflex is guided by a molecular-similarity based algorithm, and FlexX employs an incremental construction algorithm as a search engine [2]. They then combined the generated poses of each program and selected a subset of 100 decoys according to a systematic clustering procedure that will be explained in more detail in Sect. 2.3. The intention behind using four different docking algorithms was to explore the conformational space as thoroughly as possible and to avoid a potential sampling bias of this space if only one program were to be used.

In previous work, we have presented BA-based ML models for the ligand scoring and ranking problems [3,4]. However, the focus of this work is on the ligand docking problem and we present docking-specialized ML SFs in which we consider a more diverse collection of features and an explicit modeling of RMSD of binding poses, which dramatically improve docking performance.

## 1.3  Key Contributions

Various nonparametric ML methods inspired from statistical learning theory are examined in this work to model the unknown function that maps structural and physicochemical information of a protein-ligand complex to a corresponding distance to the native pose (in terms of RMSD value). Ours is the first work to perform a comprehensive assessment of the docking accuracies of conventional and machine-learning (ML) SFs across both diverse and homogeneous (protein-family-specific) test sets using a common diverse set of features across the ML SFs. We show that the best ML SF has a success rate of ∼80 % compared to ∼70 % for the best conventional SF when the goal is to find poses within RMSD of 1 Å from the native ones for 195 different protein-ligand complexes. Such a significant improvement ($> 14 \%$) in docking power will lead to better quality drug hits and ultimately help reduce costs associated with drug discovery.

We seek to advance structure-based drug design by designing SFs that significantly improve upon the protein-ligand modeling performance of conventional SFs. Our approach is to couple the modeling power of flexible machine learning algorithms with training datasets comprising hundreds of protein-ligand complexes with native poses of known high-resolution 3D crystal structures and experimentally-determined binding affinities. In addition, we computationally generate a large number of decoy poses and utilize their RMSD values from the native pose and a variety of features characterizing each complex. We will compare the docking accuracies of several ML and existing conventional SFs of all three types, force-field, empirical, and knowledge-based, on diverse and independent test sets. Further, we assess the impact of training set size on the docking performance of the conventional BA-based SFs and the proposed RMSD-based models.

The remainder of the paper is organized as follows. Section 2 presents the compound database used for the comparative assessment of SFs (Sect. 2.1), the physicochemical features extracted to characterize the compounds (Sect. 2.2), the procedure for decoy generation and formation of training and test datasets (Sect. 2.3), and conventional SFs (Sect. 2.4) and the ML methods (Sect. 2.5) that we employ. Next, in Sect. 3, we present results comparing the docking powers of conventional and ML SFs on diverse (Sect. 3.2) and homogeneous (Sect. 3.3) test sets, and analyze how they are impacted by training set size (Sect. 3.4). Finally, we close with concluding remarks in Sect. 4.

## 2   Materials and Methods

### 2.1   Compound Database

We used the 2007 version of PDBbind [16], the same complex database that Cheng et al. used as a benchmark in their recent comparative assessment of sixteen popular conventional SFs [2]. PDBbind is a selective compilation of the Protein Data Bank (PDB) database [17]. Both databases are publicly accessible and regularly updated. The PDB is periodically mined and only complexes that are suitable for drug discovery are filtered into the PDBbind database. In PDBbind, a number of filters are imposed to obtain high-quality protein-ligand complexes with both experimentally-determined BA and three-dimensional structure from PDB [2]. A total of 1300 protein-ligand complexes are compiled into a *refined set* after applying rigorous and systematic filtering criteria. The PDBbind curators compiled another list out of the refined set. It is called the *core set* and is mainly intended to be used for benchmarking docking and scoring systems. The core set is composed of diverse protein families and diverse binding affinities. BLAST [18] was employed to cluster the refined set based on protein sequence similarity with a 90 % cutoff. From each resultant cluster, three protein-ligand complexes were selected to be its representatives in the core set. A cluster must fulfill the following criteria to be admitted into the core set: (i) it has at least four members and (ii) the BA of the highest-affinity complex must be at least 100-fold of that of the complex with the lowest one. The representatives were then chosen based on their BA rank: the complex having the highest rank, the middle one, and the one with the lowest rank. The approach of constructing the core set guarantees unbiased, reliable, and biochemically rich test set of complexes. In order to be consistent with the comparative framework used to assess the sixteen conventional SFs mentioned above [2], we too consider the 2007 version of PDBbind which consists of a 1300-complex refined set and a 195-complex core set (with 65 clusters).

## 2.2   Compound Characterization

For each protein-ligand complex, we extracted physicochemical features used in the empirical SFs X-Score [6] (a set of 6 features denoted by $X$) and Aff-iScore [19] (a set of 30 features denoted by $A$) and calculated by GOLD [9] (a set of 14 features denoted by $G$), and geometrical features used in the ML SF RF-Score [20] (a 36-feature set denoted by $R$). The software packages that calculate X-Score, AffiScore (from SLIDE), and RF-Score features were available to us in an open-source form from their authors and a full list of these features are provided in the appendix of [4]. The GOLD docking suite provides a utility that calculates a set of general descriptors for both molecules. The set includes some common ligand molecular properties such as: molecular weight, number of rotatable bonds, number of hydrogen bonds, solvent exposed descriptors, etc. Protein-specific features are also calculated that account for the number of polar, acceptor, and donatable atoms buried in the binding pocket. As a complex, two protein-ligand interaction features are calculated which are the number of ligand atoms forming H-bonds and the number of ligand atoms that clash with protein atoms. The full set of these features can be easily accessed and calculated via the *Descriptors* menu in GOLD.

## 2.3   Decoy Generation and Formation of Training and Test Sets

The training dataset derived from the 2007 refined set is referred to as the *primary training set* (1105 complexes) and we denote it by $Pr$. It is composed of the 1300 refined-set complexes of 2007, excluding those 195 complexes present in the core set of the same year's version. The proteins of both these sets form complexes with ligands that were observed bound to them during 3D structure identification. These ligands are commonly known as native ligands and the conformation in which they were found at their respective binding sites are referred to as true or native poses. In order to assess the docking power of SFs in distinguishing true poses from random ones, a decoy set was generated for each protein-ligand complex in $Pr$ and $Cr$. We utilize the decoy set produced for the core set $Cr$ by Cheng et al. [2] using four popular docking tools: LigandFit in Discovery Studio, Surflex in SYBYL, FlexX in SYBYL (currently in LeadIT [21]), and GOLD. From each tool, a diverse set of binding poses was generated by controlling docking parameters as described in [2]. This process generated a total of ~2000 poses for each protein-ligand complex from the four docking protocols combined. Binding poses that are more than 10 Å away, in terms of RMSD (root-mean-square deviation), from the native pose are discarded. The remaining poses are then grouped into ten 1 Å bins based on their RMSD values from the native binding pose. Binding poses within each bin were further clustered into ten clusters based on their similarities [2]. From each such subcluster, the pose with the lowest noncovalent interaction energy with the protein was selected as a representative of that cluster and the remaining poses in that cluster were discarded. Therefore, at the end of this process, decoy sets consisting of (10 bins × 10 representatives =) 100 diverse poses were generated for each protein-ligand complex. Since we have access to the original $Cr$ decoy set, we used it as is and we followed the same procedure to generate the decoy set for the training data $Pr$. Since we did not have access to Discovery Studio software, we did not use LigandFit protocol for the training data. In order to keep the size of the training set reasonable, we generated 50 decoys for each protein-ligand complex instead of 100 as it is the case for $Cr$ complexes. Due to geometrical constraints during decoy generation, the final number of resultant decoys for some complexes does not add up exactly to 50 for $Pr$ and 100 for $Cr$. It should be noted that the decoys in the training set are completely independent of those in the test set since both datasets share no ligands from which these decoys are generated.

We develop two types of ML SFs in this work. The first type are trained to predict binding affinities (BAs) and use these scores to distinguish promising poses from less promising ones. The second set involves building SFs to predict RMSD values explicitly. As it will be shown later, this novel approach has a superior accuracy over conventional BA-based prediction. Accordingly, two versions of training and test data sets are created. The first version uses BA as the dependent variable ($Y = $ BA) and the size of $Pr$ remains fixed at 1,105 while $Cr$ includes 16,554 complexes because it consists of native poses and a decoy set for each pose. The dependent variable of the second version is RMSD ($Y = $ RMSD) and because both training and test sets consist of native and decoy poses, the size of $Pr$ expands to 39,085 while $Cr$ still retains the 16,554 complex conformations.

For all protein-ligand complexes, for both native poses and computationally-generated decoys, we extracted $X$, $A$, $R$, and $G$ features. By considering all fifteen combinations of these four types of features (i.e., $X$, $A$, $R$, $G$, $X \cup A$, $X \cup R$, $X \cup G$, $A \cup R$, $A \cup G$, $R \cup G$, $X \cup A \cup R$, $X \cup A \cup G$, $X \cup R \cup G$, $A \cup R \cup G$, and $X \cup A \cup R \cup G$), we generated ($15 \times 2 =$) 30 versions of the $Pr$ and $Cr$ data sets, which we distinguish by using the notation $Pr_F^Y$ and $Cr_F^Y$ to denote that the data set is characterized by the feature set $F$ and its dependent variable is $Y$. For instance, $Pr_{XR}^{BA}$ denotes the version of $Pr$ comprising the set of features $X \cup R$ (referred to simply as XR) and experimentally-determined BA data for complexes in the $Pr$ dataset.

## 2.4 Conventional Scoring Functions

A total of sixteen popular conventional SFs are compared to ML SFs in this study. The sixteen functions are either used in mainstream commercial docking tools and/or have been developed in academia. The functions were recently compared against each other in a study conducted by Cheng et al. [2]. This set includes five SFs in the Discovery Studio software [11]: LigScore, PLP, PMF, Jain, and LUDI. Five SFs in SYBYL software [22]: D-Score, PMF-Score, G-Score, ChemScore, and F-Score. GOLD software [9] contributes three SFs: GoldScore, ChemScore, and ASP. GlideScore in the Schrödinger software [23]. Besides, two standalone scoring functions developed in academia are also assessed, namely, DrugScore [12] and X-Score [6]. Some of the SFs have several options or versions, these include LigScore (LigScore1 and LigScore2), PLP (PLP1 and PLP2), and LUDI (LUDI1, LUDI2, and LUDI3) in Discovery Studio; GlideScore (GlideScore-SP and GlideScore-XP) in the Schrödinger software; DrugScore (DrugScore-PDB and DrugScore-CSD); and X-Score (HPScore, HMScore, and HSScore). For brevity, we only report the version and/or option that yields the best performance on the PDBbind benchmark that was considered by Cheng et al.

## 2.5   Machine Learning Methods

We utilize a total of six regression techniques in our study: multiple linear regression (MLR), multivariate adaptive regression splines (MARS), $k$-nearest neighbors ($k$NN), support vector machines (SVM), random forests (RF), and boosted regression trees (BRT) [24]. These techniques are implemented in the following R language packages that we use [25]: the package *stats* readily available in R for MLR, *earth* for MARS [26], *kknn* for $k$NN [27], *e1071* for SVM [28], *randomForest* for RF [29], and *gbm* for BRT [30]. These methods benefit from some form of parameter tuning prior to their use in prediction. The optimal parameter values we use to build our models resulted from a grid search associated with 10-fold cross validation over the training set $Pr$ and are provided in [4]. These values are obtained based on $Pr_F^{BA}$ for any given feature set $F$; optimizing based on $Pr_F^{RMSD}$ yielded similar parameter values, therefore, for brevity, we do not include them here. For every machine-learning method, we will be using these values to build ML SFs in the subsequent experiments.

# 3   Results and Discussion

## 3.1   Evaluation of Scoring Functions

In contrast to our earlier work in improving and examining scoring and ranking accuracies of different families of SFs [3,4], this study is devoted to enhancing and comparing SFs in terms of their docking powers. Docking power measures the ability of an SF to distinguish a promising binding mode from a less promising one. Typically, generated conformations are ranked in non-ascending order according to their predicted binding affinity (BA). Ligand poses that are very close to the experimentally-determined ones should be ranked high. Closeness is measured in terms of RMSD (in Å) from the true binding pose. Generally, in docking, a pose whose RMSD is within 2 Å from the true pose is considered a success or a hit.

In this work, we use comparison criteria similar to those used by Cheng et al. to compare the docking accuracies of sixteen popular conventional SFs. Doing so ensures fair comparison of ML SFs to those examined in that study in which each SF was assessed in terms of its ability to find the pose that is closest to the native one. More specifically, docking ability is expressed in terms of a success rate statistic $S$ that accounts for the percentage of times an SF is able to find a pose whose RMSD is within a predefined cutoff value $C$ Å by only considering the $N$ topmost poses ranked by their predicted scores. Since success rates for various $C$ (e.g., 0, 1, 2, and 3 Å) and $N$ (e.g., 1, 2, 3, and 5) values are reported in this study, we use the notation $S_C^N$ to distinguish between these different statistics. For example, $S_1^2$ is the percentage of protein-ligand complexes whose either one of the two best scoring poses are within 1 Å from the true pose of a given complex. It should be noted that $S_0^1$ is the most stringent docking measure in which an SF is considered successful only if the best scoring pose is the native pose. By the same token and based on the $C$ and $N$ values listed earlier, the least strict docking performance statistic is $S_3^5$ in which an SF is considered successful if at least one of the five best scoring poses is within 3 Å from the true pose.

## 3.2   ML vs. Conventional Approaches on a Diverse Test Set

After building six ML SFs, we compare their docking performance to the sixteen conventional SFs on the core test $Cr$ that comprises thousands of protein-ligand complex conformations corresponding to 195 different native poses in 65 diverse protein families. As mentioned earlier, we conducted two experiments. In the first, BA values predicted using the conventional and ML SFs were used to rank poses in a non-ascending order for each complex in $Cr$. In the other experiment, RMSD-based ML models directly predicted RMSD values that are used to rank in non-descending order the poses for the given complex.

By examining the true RMSD values of the best $N$ scoring ligands using the two prediction approaches, success rates of SFs are shown in Fig. 2. Panels (a) and (b) in the figure show the success rates $S_1^1$, $S_2^1$, and $S_3^1$ for all 22 SFs. The SFs, as in the other panels, are sorted in non-ascending order from the most stringent docking test statistic value to the least stringent one. In the top two panels, for example, success rates are ranked based on $S_1^1$, then $S_2^1$ in case of a tie in $S_1^1$, and finally $S_3^1$ if two or more SFs tie in $S_2^1$. In both BA- and RMSD-based scoring, we find that the 22 SFs vary significantly in their docking performance. The top three BA-based SFs, GOLD::ASP, DS::PLP1, and DrugScorePDB::PairSurf, have success rates of more than 60 % in terms of $S_1^1$ measure. That is in comparison to the BA-based ML SFs, the best of which has an $S_1^1$ value barely exceeding 50 % (Fig. 2(a)). On the other hand, the other six ML SFs that directly predict RMSD values achieve success rates of over 70 % as shown in Fig. 2(b). The top performing of these ML SFs, MARS::XARG, has a success rate of $\sim$80 %. This is a significant improvement ($>$14 %) over the best conventional SF, the empirical GOLD::ASP, whose $S_1^1$ value is $\sim$70 %. Similar conclusions can also be made for the less stringent docking performance measures $S_2^1$ and $S_3^1$ in which the RMSD cut-off constraint is relaxed to 2 Å and 3 Å, respectively.

The success rates plotted in the top two panels (Fig. 2 (a) and (b)) are reported when native poses are included in the decoy sets. Panels (c) and (d) of the same figure show the impact of removing the native poses on docking success rates of all SFs. It is clear that the performance of almost all SFs does not radically decrease by examining the difference in their $S_2^1$ statistics which ranges from 0 to $\sim$5 %. This, as it was noted by Cheng et al. [2], is due to the fact that

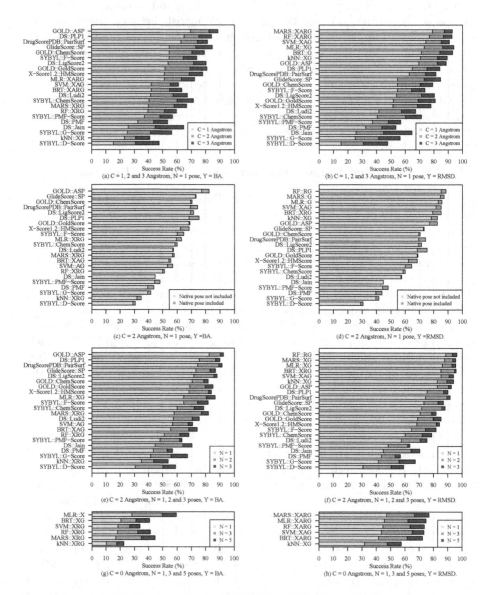

**Fig. 2.** Success rates of conventional and ML SFs in identifying binding poses that are closest to native ones. The results show these rates by examining the top $N$ scoring ligands that lie within an RMSD cut-off of $C$ Å from their respective native poses. Panels on the left show success rates when binding-affinity based ($BA$) scoring is used and the ones on the right show the same results when ML SFs predicted $RMSD$ values directly. Scoring of conventional SFs is BA-based in all cases and for comparison convenience we show their performance in the right panels as well.

some of the poses in the decoy sets are actually very close to the native ones. As a result, the impact of allowing native poses in the decoy sets is insignificant in most cases and therefore we include such poses in all other tests in the paper.

In reality, more than one pose is usually used from the outcomes of a docking run in the next stages of drug design for further experimentation. It is useful therefore to assess docking accuracy of SFs when more than one pose is considered (i.e., $N > 1$). Figure 2 (e) and (f) show the success rates of SFs when the RMSD values of the best 1, 2, and 3 scoring poses are examined. These rates correspond, respectively, to $S_1^2$, $S_2^2$, and $S_3^2$. The plots show a significant boost in performance for almost all SFs. By comparing $S_1^2$ to $S_3^2$, we observe a jump in accuracy from 82 % to 92 % for GOLD::ASP and from 87 % to 96 % for RF::RG that models RMSD values directly. Such results signify the importance of examining an ensemble of top scoring poses because there is a very good chance it contains relevant conformations and hence good drug candidates.

Upon developing RMSD-based ML scoring models, we noticed excellent improvement over their binding-affinity-based counterparts as shown in Fig. 2. We conducted an experiment to investigate whether they will maintain a similar level of accuracy when ML SFs are examined for their ability to pinpoint the native poses from their respective 100-pose decoy sets. The bottom two panels, (g) and (h), plot the success rates in terms of $S_0^1$, $S_0^3$, and $S_0^5$ for the six ML SFs. By examining the five best scoring poses, we notice that the top BA-based SF, MLR::X, was able to distinguish native binding poses in ∼60 % of the 195 decoy sets whereas the top RMSD-based SF, MARS::XARG, achieved a success rate of $S_0^5 = 77$ % on the same protein-ligand complexes. It should be noted that both sets of ML SFs, the BA- and RMSD-based, were trained and tested on completely disjoint test sets. Therefore, this gap in performance is largely due to the explicit modeling of RMSD values and the corresponding abundance of training data which includes information from both native and computationally-generated poses.

### 3.3   ML vs. Conventional Approaches on Homogeneous Test Sets

In the previous section, performance of SFs was assessed on the diverse test set $Cr$. The core set consists of more than sixty different protein families each of which is related to a subset of protein families in $Pr$. That is, while the training and test set complexes were different (at least for all the ML SFs), proteins present in the core test set were also present in the training set, albeit bound to different ligands. A much more stringent test of SFs is their evaluation on a completely new protein, i.e., when test set complexes all feature a given protein—test set is homogeneous—and training set complexes do not feature that protein. To address this issue, four homogeneous test sets were constructed corresponding to the four most frequently occurring proteins in our data: HIV protease (112 complexes), trypsin (73), carbonic anhydrase (44), and thrombin (38). Each of these protein-specific test sets was formed by extracting complexes containing the protein from $Cr$ (one cluster or three complexes) and $Pr$ (remaining complexes). For each test set, we retrained BRT, RF, SVM, $k$NN, MARS, and MLR models on the non-test-set complexes of $Pr$. Figure 3 shows the docking performance of resultant BA and RMSD-based ML scoring models on the four protein families. The plots clearly show that success rates of SFs are dependent on the protein family under investigation. It is easier for some SFs to distinguish good poses for HIV protease and thrombin than for carbonic anhydrase. The best performing SFs on HIV protease and thrombin complexes, MLR::XRG and MLR::XG, respectively, achieve success rates of over 95 % in terms of $S_1^3$ as shown in panels (b) and (n), whereas no SF exceeded 65 % in success rate in case of carbonic anhydrase as demonstrated in panels (i) and (j). Finding the native poses is even more challenging for all SFs, although we can notice that RMSD-based SFs outperform those models that rank poses using predicted BA. The exception to this is the SF MLR::XAR whose performance exceeds all RMSD-based ML models in terms of the success rate in reproducing native poses as illustrated in panels (c) and (d).

The results also indicate that multivariate linear regression models (MLR), which are basically empirical SFs, are the most accurate across the four families, whereas ensemble learning models, RF and BRT, unlike their good performance in Fig. 2, appear to be inferior compared to simpler models in Fig. 3. This can be attributed to the high rigidity of linear models compared to ensemble approaches. In other words, linear models are not as sensitive as ensemble techniques to the presence or absence of certain protein family in the data on which they are trained. On the other hand, RF- and BRT-based SFs are more flexible and adaptive to their training data that in some cases fail to generalize well enough to completely different test proteins as seen in Fig. 3. In practice, however, it has been observed that more than 92 % of today's drug targets are similar to known proteins in PDB [31], an archive of high quality complexes from which our training and test compounds originated. Therefore, if the goal of a docking run is to identify the most stable poses, it is important to consider sophisticated SFs (such as RF and BRT) calibrated with training sets containing some known binders to the target of interest. Simpler models, such as MLR and MARS, tend to be more accurate when docking to novel proteins that are not present in training data.

Sophisticated ML algorithms are not the only critical element in building a capable SF. Features to which they are fitted also play an important role as can be seen in Fig. 3. By comparing the right panels to the ones on the left, we can notice that X-Score features (X) are almost always present in BA-based SFs while those provided by GOLD (G) are used more to model RMSD explicitly. This implies that X-Score features are more accurate than other feature sets in predicting BA, while GOLD features are the best for estimating RMSD and hence poses close to the native one.

## 3.4    Impact of Training Set Size

An important factor influencing the accuracy of ML SFs is the size of the training dataset. In the case of BA-based ML SFs, training dataset size can be increased by training on a larger set of protein-ligand complexes with known

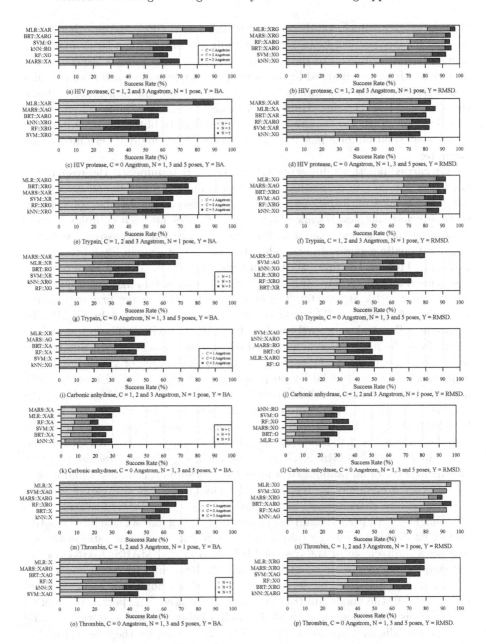

**Fig. 3.** Success rates of ML SFs in identifying binding poses that are closest to native ones observed in four protein families: HIV protease (a-d), trypsin (e-h), carbonic anhydrase (i-l), and thrombin (m-p). The results show these rates by examining the top $N$ scoring ligands that lie within an RMSD cut-off of $C$ Å from their respective native poses. Panels on the left show success rates when binding-affinity based ($BA$) scoring is used and the ones on the right show the same results when ML SFs predicted $RMSD$ values directly.

binding affinity values. In the case of RMSD-based SFs, on the other hand, training dataset size can be increased not only by considering a large number of protein-ligand complexes in the training set, but also by using a larger number of computationally-generated ligand poses per complex since each pose provides a new training record because it corresponds to a different combination of features and/or RMSD value. Unlike experimental binding affinity values, which have inherent noise and require additional resources to obtain, RMSD from the native conformation for a new ligand pose is computationally determined and is accurate.

We carried out three different experiments to determine: (i) the response of BA-based ML SFs to increasing number of training protein-ligand complexes, (ii) the response of RMSD-based ML SFs to increasing number of training protein-ligand complexes while the number of poses for each complex is fixed at 50, and (iii) the response of RMSD-based ML SFs to increasing number of computationally-generated poses while the number of protein-ligand complexes is fixed at 1105. In the first two experiments, we built 6 ML SFs, each of which was trained on a randomly sampled $x\,\%$ of the 1105 protein-ligand complexes in $Pr$, where $x = 10, 20, \ldots, 100$. The dependent variable in the first experiment is binding affinity ($Y = BA$), and the performance of these BA-based ML SFs is shown in Fig. 4(a) and partly in Fig. 4(d) (MLR::XARG). The set of RMSD values from the native pose is used as a dependent variable for ML SFs trained in the second experiment ($Y = RMSD$). For a given value of $x$, the number of conformations is fixed at 50 ligand poses for each protein-ligand complex. The docking accuracy of these RMSD-based ML models is shown in Fig. 4(b). In the third experiment, all 1105 complexes in $Pr$ were used for training the RMSD-based ML SFs (i.e., $Y = RMSD$) with $x$ randomly sampled poses considered per complex, where $x = 2, 6, 10, \ldots, 50$; results for this are reported in Fig. 4(c) and partly in Fig. 4(d) (MARS::XARG). In all three experiments, results reported are the average of 50 random runs in order to ensure all complexes and a variety of poses are equally represented. All training and test complexes in these experiments are characterized by the XARG ($=X \cup A \cup R \cup G$) features.

From Fig. 4(a), it is evident that increasing training dataset size has a positive impact on docking accuracy (measured in terms of $S_1^1$ success rate), although it is most appreciable in the case of MLR::XARG and MARS::XARG, two of the simpler models, MLR being linear and MARS being piecewise linear. The performance of the other models, which are all highly nonlinear, seems to saturate at 60 % of the maximum training dataset size used. The performance of all six models is quite modest, with MLR::XARG being the only one with docking success rate (slightly) in excess of 50 %. The explanation for these results is that binding affinity is not a very good response variable to learn for the docking problem because the models are trained only on native poses (for which binding affinity data is available) although they need to be able to distinguish between native and non-native poses during testing. This means that the training data is not particularly well suited for the task for which these models are used. An additional reason is that experimental binding affinity data, though useful,

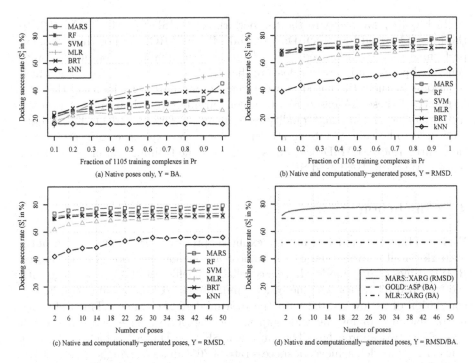

**Fig. 4.** Dependence of docking accuracy of ML scoring models on training set size when training complexes are selected randomly (without replacement) from $Pr$ and the models are tested on $Cr$. The size of the training data was increased by including more protein-ligand complexes ((a) and (b)) or more computationally-generated poses for all complexes ((c) and (d)).

is inherently noisy. The flexible highly nonlinear models, RF, BRT, SVM, and $k$NN, are susceptible to this noise because the training dataset (arising only from native poses) is not particularly relevant to the test scenario (consisting of both native and non-native poses). Therefore, the more rigid MLR and MARS models fair better in this case.

When RMSD is used as the response variable, the training set consists of data from both native and non-native poses and hence is more relevant to the test scenario and the RMSD values, being computationally determined, are also accurate. Consequently, docking accuracy of all SFs improves dramatically compared to their BA-based counterparts as can be observed by comparing Fig. 4(a) to Fig. 4(b) and (c). We also notice that all SFs respond favorably to increasing training set size by either considering more training complexes (Fig. 4(b)) or more computationally-generated training poses (Fig. 4(c)). Even for the smallest training set sizes in Fig. 4(b) and (c), we notice that the docking accuracy of most RMSD-based SFs is about 70 % or more, which is far better than the roughly 50 % success rate for the largest training set size for the best BA-based SF MLR::XARG.

In Fig. 4(d), we compare the top performing RMSD SF, MARS::XARG, to the best BA-based SFs, GOLD::ASP and MLR::XARG, to show how docking performance can be improved by just increasing the number of computationally-generated poses, an important feature that RMSD-based SFs possess but which is lacking in their BA-based conventional counterparts. To increase the performance of these BA-based SFs to a comparable level, thousands of protein-ligand complexes with high-quality experimentally-determined binding affinity data need to be collected. Such a requirement is too expensive to meet in practice. Furthermore, RMSD-based SFs with the same training complexes will still likely outperform BA-based SFs.

## 4   Conclusion

We found that ML models trained to explicitly predict RMSD values significantly outperform all conventional SFs in almost all testing scenarios. The estimated RMSD values of such models have a correlation coefficient of 0.7 on average with the true RMSD values. On the other hand, predicted binding affinities have a correlation of as low as -0.2 with the measured RMSD values. This difference in correlation explains the wide gap in docking performance between the top SFs of the two approaches. The empirical SF GOLD::ASP, which is the best conventional model, achieved a success rate of 70 % in identifying a pose that lies within 1 Å from the native pose of 195 different complexes. On other hand, our top RMSD-based SF, MARS::XARG, has a success rate of ~80 % on the same test set, which represents a significant improvement in docking performance. We also observed steady gains in the performance of RMSD-based ML SFs as the training set size was increased by considering more protein-ligand complexes and/or more computationally-generated ligand poses for each complex.

## References

1. Lyne, P.D.: Structure-based virtual screening: an overview. Drug Discov. Today **7**(20), 1047–1055 (2002)
2. Cheng, T., Li, X., Li, Y., Liu, Z., Wang, R.: Comparative assessment of scoring functions on a diverse test set. J. Chem. Inf. Model. **49**(4), 1079–1093 (2009)
3. Ashtawy, H.M., Mahapatra, N.R.: A comparative assessment of conventional and machine-learning-based scoring functions in predicting binding affinities of protein-ligand complexes. In: 2011 IEEE International Conference on Bioinformatics and Biomedicine (BIBM), pp. 627–630. IEEE (2011)
4. Ashtawy, H.M., Mahapatra, N.R.: A comparative assessment of ranking accuracies of conventional and machine-learning-based scoring functions for protein-ligand binding affinity prediction. IEEE/ACM Trans. Comput. Biol. Bioinf. (TCBB) **9**(5), 1301–1313 (2012)
5. Ewing, T., Makino, S., Skillman, A., Kuntz, I.: Dock 4.0: search strategies for automated molecular docking of flexible molecule databases. J. Comput. Aided Mol. Des. **15**(5), 411–428 (2001)

6. Wang, R., Lai, L., Wang, S.: Further development and validation of empirical scoring functions for structure-based binding affinity prediction. J. Comput. Aided Mol. Des. **16**, 11–26 (2002). doi:10.1023/A:1016357811882

7. Gohlke, H., Hendlich, M., Klebe, G.: Knowledge-based scoring function to predict protein-ligand interactions. J. Mol. Biol. **295**(2), 337–356 (2000)

8. Mooij, W., Verdonk, M.: General and targeted statistical potentials for protein-ligand interactions. Proteins **61**(2), 272 (2005)

9. Jones, G., Willett, P., Glen, R., Leach, A., Taylor, R.: Development and validation of a genetic algorithm for flexible docking. J. Mol. Biol. **267**(3), 727–748 (1997)

10. Gehlhaar, D.K., Verkhivker, G.M., Rejto, P.A., Sherman, C.J., Fogel, D.R., Fogel, L.J., Freer, S.T.: Molecular recognition of the inhibitor AG-1343 by HIV-1 protease: conformationally flexible docking by evolutionary programming. Chem. Biol. **2**(5), 317–324 (1995)

11. Inc., A.S.: The Discovery Studio Software, San Diego, CA (2001) (version 2.0)

12. Velec, H.F.G., Gohlke, H., Klebe, G.: DrugScore CSD - knowledge-based scoring function derived from small molecule crystal data with superior recognition rate of near-native ligand poses and better affinity prediction. J. Med. Chem. **48**(20), 6296–6303 (2005)

13. Venkatachalam, C., Jiang, X., Oldfield, T., Waldman, M.: LigandFit: a novel method for the shape-directed rapid docking of ligands to protein active sites. J. Mol. Graph. Model. **21**(4), 289–307 (2003)

14. Jain, A.: Surflex-dock 2.1: robust performance from ligand energetic modeling, ring flexibility, and knowledge-based search. J. Comput. Aided Mol. Des. **21**(5), 281–306 (2007)

15. Rarey, M., Kramer, B., Lengauer, T., Klebe, G.: A fast flexible docking method using an incremental construction algorithm. J. Mol. Biol. **261**(3), 470–489 (1996)

16. Wang, R., Fang, X., Lu, Y., Wang, S.: The PDBbind database: collection of binding affinities for protein-ligand complexes with known three-dimensional structures. J. Med. Chem. **47**(12), 2977–2980 (2004)

17. Berman, H.M., Westbrook, J., Feng, Z., Gilliland, G., Bhat, T.N., Weissig, H., Shindyalov, I.N., Bourne, P.E.: The protein data bank. Nucleic Acids Res. **28**(1), 235–242 (2000)

18. Madden, T.: The BLAST sequence analysis tool. In: McEntyre, J., Ostell, J. (eds.) The NCBI Handbook. National Library of Medicine (US), National Center for Biotechnology Information, Bethesda (2002)

19. Schnecke, V., Kuhn, L.A.: Virtual screening with solvation and ligand-induced complementarity. In: Klebe, G. (ed.) Virtual Screening: An Alternative or Complement to High Throughput Screening?, pp. 171–190. Springer, Amsterdam (2002)

20. Ballester, P., Mitchell, J.: A machine learning approach to predicting protein-ligand binding affinity with applications to molecular docking. Bioinformatics **26**(9), 1169 (2010)

21. BioSolveIT.: LeadIT, St. Augustin, Germany (2012) (version 2.1)

22. Inc., T.: The SYBYL Software, 1699 South Hanley Rd., St. Louis, Missouri, 63144, USA (2006) (version 7.2)

23. Schrödinger, L.: The Schrödinger Software, New York (2005) (version 8.0)

24. Hastie, T., Tibshirani, R., Friedman, J.: The Elements of Statistical Learning. Springer, New York (2001)

25. Team, R.D.C.: R: A Language and Environment for Statistical Computing. R Foundation for Statistical Computing, Vienna, Austria (2010) ISBN 3-900051-07-0

26. Milborrow, S., Trevor, H., Tibshirani, R.: earth: Multivariate Adaptive Regression Spline Models (2010) (R package version 2.4-5)

27. Hechenbichler, K.S.K.: kknn: Weighted k-Nearest Neighbors (2010) (R package version 1.0-8)
28. Dimitriadou, E., Hornik, K., Leisch, F., Meyer, D., Weingessel, A.: e1071: Miscellaneous Functions of the Department of Statistics (e1071), TU Wien (2010) (R package version 1.5-24)
29. Breiman, L.: Random forests. Mach. Learn. **45**, 5–32 (2001)
30. Ridgeway, G.: gbm: Generalized Boosted Regression Models (2010) (R package version 1.6-3.1)
31. Overington, J., Al-Lazikani, B., Hopkins, A.: How many drug targets are there? Nat. Rev. Drug Discovery **5**(12), 993–996 (2006)

# BioCloud Search EnGene:
# Surfing Biological Data on the Cloud

Nicoletta Dessì[✉], Emanuele Pascariello, Gabriele Milia,
and Barbara Pes

Dipartimento di Matematica e Informatica,
Università degli Studi di Cagliari, Via Ospedale 72, 09124 Cagliari, Italy
{dessi,milia.ga,pes}@unica.it,
emanuele.pascariello@gmail.com

**Abstract.** The massive production and spread of biomedical data around the web introduces new challenges related to identify computational approaches for providing quality search and browsing of web resources. This papers presents BioCloud Search EnGene (BSE), a cloud application that facilitates searching and integration of the many layers of biological information offered by public large-scale genomic repositories. Grounding on the concept of dataspace, BSE is built on top of a cloud platform that severely curtails issues associated with scalability and performance. Like popular online gene portals, BSE adopts a gene-centric approach: researchers can find their information of interest by means of a simple "Google-like" query interface that accepts standard gene identification as keywords. We present BSE architecture and functionality and discuss how our strategies contribute to successfully tackle big data problems in querying gene-based web resources. BSE is publically available at: http://biocloud-unica.appspot.com/.

**Keywords:** Biomedical data exploration · Cloud computing · Data searching · Data integration · Dataspaces · Pay-as-you-go data querying

## 1 Introduction

The massive production and spread of biomedical data around the web introduces new challenges related to identify computational approaches for their management and exploitation. These challenges mainly result from three issues:

- *Biomedical data are typical of the category of "big data"* [1]. The term "big data" refers to "the ever increasing amount of information that organizations are storing, processing and analyzing, owning the growing number of information sources in use" [2].
- *Biomedical data relay with a wide range of types and sources.* As biomedical research became interdisciplinary, information searching often requires the integration of information with multiple levels of granularities and relates data that pertain to different disciplines. Hence, the user search is not limited to a single source, but it is carried out through separate web resources in which information is represented in a different way.

© Springer International Publishing Switzerland 2014
E. Formenti et al. (Eds.): CIBB 2013, LNBI 8452, pp. 33–48, 2014.
DOI: 10.1007/978-3-319-09042-9_3

– *Biomedical data must be accessed quickly* to determine which information to show to a user on a webpage. To do global analysis, biological researchers often need to access data from multiple archival databases.

It has been observed [3] that gene sequencing technologies have become more and more affordable but the challenge of integrating disparate resources of biologic information remains difficult and more implicit or automatic ways of joining information are needed to improve the usability of gene annotation resources where searching is often unwieldy.

The development of efficient, optimized, and highly scalable search tools is a particularly challenging task as data are reaching tsunami proportions [4] and related clinical applications are seen as a "slowly rising tide" [5].

In this work we focus on genomics, a key area of biology which places greater stress on trying to solve the problem of collecting and processing large volumes of biological information, due to the fact that biological data accumulate at an ever-faster pace.

Specifically, we envision searching genetic information in databases and web resources to be like searching information in the web: we search for the information we exactly need and capture a lot of information in a short time from different websites. To face the challenge of supporting scientists in searching genetic information, we stop thinking in terms of capabilities of individual web resources and instead think of the computational functionalities needed.

In order to avoid browsing web resources and data locked to specific infrastructures, we propose advanced search functionalities on many resources via high quality, interoperable services offered in a "neutral" territory. As it happens for web engines which are designed to search for information on the World Wide Web, these services act as specialists which mine data available in many databases or open directories and return real-time information. They are a mean of organizing and integrating information from different web sources and making them manageable and satisfactory for the user.

In this article, we present BioCloud Search EnGene (BSE), a comprehensive searching environment which facilitates the versatile integration of existing genetic and genomic information from multiple heterogeneous resources. It proposes a new operational framework in which genetic information and computing technologies are reshaping each other. Like popular online gene portals, BSE adopts a gene-centric approach: researchers can find their information of interest by means of a simple "Google-like" query interface that accepts standard gene identification as keywords. Moreover, by using advanced searching and tools, users are allowed to extend their possibilities of standard data searching on popular genetic databases. BSE heavily relies on the following key design features.

First, BSE is grounded on the concept of dataspace [6, 7], a new paradigm for data integration characterized by a very loosely structured data model and intended for the management of heterogeneous data coming from a diverse set of sources regardless their format and location.

Second, to handle important coordination tasks, BSE is built on top of a cloud platform which is the physical infrastructure for hosting the dataspace. This severely curtails issues associated with scalability and performance, especially during information retrieval as searching expands across multiple server nodes. Finally, BSE is built into an integrated cloud environment that allows a close integration with web servers and standard protocols and facilitates rapid development and updates.

The paper is organized as follows. Section 2 provides background concepts and motivates the adoption of dataspace and cloud paradigms. Section 3 details the architectural aspects of BSE. The system functionalities are described in Sect. 4. Finally, Sect. 5 presents conclusions.

## 2 Background and Motivations

Dozens of gene annotation resources and databases exist which serve prominent roles in the genetics and genomics communities, each presenting a particular aspect of available gene notations. For example, the 20th annual Database Issue of Nucleic Acids Research (NAR) includes 176 articles half of which describe new online molecular biology databases and the other half provide updates on the databases previously featured in NAR and other journals.

As notable example, Entrez [8, 9] is the most popular system for searching and retrieving information from databases that are maintained by the NCBI (National Center for Biotechnology Information) [10]. Entrez is constantly being developed and improved. It indexes records in NCBI databases by means of nodes that correspond to specific databases including GenBank [11, 12], Protein database [13] and also scientific abstracts from the PubMed database [14, 15]. Access to these resources is provided by the graphical user interface of the NCBI Entrez system or by using NCBI Web services.

In exploring a database, researchers are not interested in exploiting the resource full content, but they just distil a huge amount of data to obtain succinct, key information about a concept. As biology encompasses many domains of knowledge, the success of their search depends on their ability in browsing large-scale information that is stored in several databases and web sites, each having its own organization, terminology and data formats. Unearthing specialized information can also be complex, time consuming and daunting as the researcher is also involved in learning and remembering the navigation paths of each specific web site. Finally, different web portals implement the same basic functionality and are often concerned with overlapping information.

For effective searching biomedical databases in the face of the growing number of bio-resources available worldwide, we have to answer three fundamental questions.

First, how to integrate structured, semi-structured and unstructured available data with diverse and sparse schemas?

Second, how to retrieve meaningful information in an easy and efficient way?

Finally, how implementing a searching infrastructure which has to scale, hence change, to meet new requirements stemming from the growth of its searching domain?

Computational solutions ranging from database to data warehouse poorly adapt to facing the above questions as:

(a) Many resources are large in size, dynamic, and physically distributed. Consequently, there is the need for mechanisms that can efficiently extract the relevant information from disparate sources on demand.

(b) The resources of interest are autonomously owned and operated. Consequently, searching strategies must be devised for obtaining the necessary information within the operational constraints imposed by the data source.

(c) Being heterogeneous in structure and content, information resources represent data according to their own schema which, implicitly or explicitly, defines its own concepts and relationships among concepts.

(d) Searching happens in different contexts and from different user perspectives. Hence, it is necessary to implement mechanisms for extracting context-dependent information.

The research community has recently proposed the concept of dataspace [6, 7] as a new scenario for structuring information relevant to a particular organization, regardless of its format and location. The elements of a dataspace are a set of participants (i.e., individual data sources) and a set of relationships between them [7]. In this sense, a dataspace is an abstraction of a database that does not require data to be structured and has a minimal "off-the-shelf" set of search functions based on keywords. The key idea is to enhance the quality of data integration and the semantic meaning of information without an a priori schema for the data sources [16–18].

In sharp contrast to the traditional approaches, a dataspace is based on a "data coexistence" approach as it integrates data according to a very loosely structured data model which is intended for the management of heterogeneous data coming from a diverse set of sources. A fundamental part of a dataspace is the catalogue that contains information about participants and their relationships with associate mechanisms for its gradually extension. Advanced DBMS-like functions, queries and mappings are provided over time by different components, each defining relationships among data when required. Integrated views over a set of data sources are provided following the so-called pay-as-you-go principle that is currently emerging on the web [19, 20].

In this work, we choose the dataspace paradigm as a data integration architecture for reconciling data from heterogeneous sources and providing users with a unified view of these data.

Our key idea is to conceive BSE as a cloud based application which essentially rents its capacity from a cloud computing platform.

Recent research [21] has proposed cloud computing as an innovative computational environment for searching large-scale data in a more efficient way. Specifically, cloud computing refers to a flexible and scalable internet infrastructure where processing and storage capability are dynamically provided. A cloud infrastructure abstracts the underlying hardware (i.e. servers, networking, storage etc.) and enables on-demand network access to a shared pool of computing resources that can be readily provisioned and released.

The next section will present how the BSE architecture benefits from both the dataspace and the cloud paradigms.

# 3  Architectural Aspects

Grounded on the dataspace paradigm, BSE undertakes the responsibility of coordinating and organizing the search across different web resources that are assumed to be the dataspace participants.

Data integration expects no data transfer to any central repository, except for the data stored in BSE catalogue which is initially built and gradually updated. In some way, this catalogue has the same role of the table of facts in a data warehouse where the dimension tables are distributed across many web resources. However, it differs from a data warehouse schema because:

(1)  It contains information about various participants instead of relational tables.
(2)  Besides storing and indexing participants, the catalogue contains mechanisms for creating new relationships by modifying the existing ones.
(3)  It avoids the definition of an a priori matching schema.

From a logical point of view, the catalogue is a multi-level index that specifies how genetic information from various web resources is captured and linked together. Physically, it is implemented by an object-oriented database, specifically a key-value NoSQL database [22], which stores gene annotations, acquires and combines information from external resources that participate in the dataspace.

The current version of BSE implements a dataspace with the contribution of 34 participants. According to their role in supplying data, these participants are categorized as:

– *Local participants*, i.e. resources from which some useful content is captured and permanently stored into the catalogue.
– *Service-based participants*, i.e. resources whose content is captured at running time by specific BSE services in a pay-as-you-go fashion according to the user request.
– *External participants*, i.e. resources whose web links are dynamically built and activated when it is required.

The catalogue organizes objects in classes, each corresponding to one local participant. Table 1 shows the list of local participants and the corresponding catalogue content.

BSE also relies on the following external services:

– NCBI Entrez Programming Utilities (E-utilities) [23]
– UniChem RESTful Web Service API [24]
– Database identifier mapping [25]
– STRING API [26]
– WikiPathways Webservice/API [27]
– REST-style version of KEGG API [28]
– mygene.info REST web services [29]
– RESTful web service Europe PMC [30]

**Table 1.** Local participants and corresponding catalogue content.

| Dataset | Catalogue content |
| --- | --- |
| Entrez Gene homo sapiens gene info [32] | Main annotations about human genes |
| Entrez Gene Relations human gene relations [32] | Gene to gene relationships |
| M.A.T.A.D.O.R. Manually annotated targets and drugs online resource [33] | Gene drug relationships |
| Entrez gene ID to Pathways [34] | Human genes pathways, according to Reactome |
| Entrez gene ID to Mendelian Phenotype [35] | Human mendelian phenotypes and their gene associations |
| Entrez gene ID to RefSeq [36] | Cumulative set of transcripts and proteins |
| H.A.G.R. - Human Ageing Genome Resource [37] | Genes possibly related to human ageing |
| Wellcome Trust Sanger Institute - cancer genomics annotations [38] | Cancer drug sensitivity annotated genes |

Conversely, a local participant is viewed as a repository of objects associated with the catalogue. Relationships between participants are expressed by means of key-values [31] that store gene identifiers as defined by international scientific standards.

### 3.1 Query Contextualization

The schema free and non-rigid structure of the catalogue allows us to implement with relative ease new ways of querying and extracting information on the basis of what we can here define as a query context or a context from now on.

Specifically, a context is a logical structure that supports queries about common points of interest the users share in browsing dataspace participants. As an example, if the user is interested in searching information about genes associated with a specific disorder, he refers to the context "Mendelian genetic disorders". Contexts are the only way to query data. Each context presents a "gene centric view" where the users can easily identify the relevant resources and navigate the content of the resources to which the context relates. Contexts hide the complexity of data searching, where BSE services capture and present the information of interest.

From a technical point of view, contexts identify specific perspectives on dataspace participants that are kept in the dataspace catalogue. These perspectives resemble to views in relational databases. However, being the catalogue implemented by a NoSQL database, they do not result from joining structured relational tables, but from relationships expressed by key-values. As well, contexts take very little space to be stored as the catalogue contains only the definition of contexts without a copy of all the data that the context relates to.

The present version of BSE implements the following contexts:

*1. Known gene name or gene identifications.*

Here, we assume that the user is able to identify genes by their standard identifier and wants to know further details.

*2. Query by Human Mendelian Genetic disorders.*

The user is allowed to extract a list of genes using the name of a certain phenotype associated with a genetic disorder with Mendelian transmission character.

*3. Query by pathway.*

This context allows the user to extract a list of human genes annotated in a given biological pathway. A pathway is a set of chemical reactions related to one or more processes within a cell. It results in expression products whose knowledge is very important in the study of biological phenomena.

*4. Bulk queries.*

This context allows to extract a list of human genes meeting the following searching criteria: gene biotype, chromosome belonging, ageing related annotation, chemotherapeutic sensitivity related to annotated genes according to their mutational status.

*5. Query by Drug information.*

It moves the query focus from a purely genetic perspective to a context dealing with the relationships between pharmacologically active molecules and the human genome expression products.

### 3.2   Technical Details

BSE is built on top of GAE (Google App Engine) [39], a platform as a service (PaaS) cloud computing environment which hosts web applications. Differently from other PaaS offerings, GAE benefits from the same infrastructure that supports basic Google applications and services such as Google search engine, You tube, Google Earth etc.

We stored the dataspace catalogue into the GAE datastore, a distributed data storage service that performs distribution, replication and load balancing automatically and supports operations to access objects (i.e. create, read, update, delete) by means of an SQL-like language called GQL.

We used Phyton as programming language and implemented BSE functionality using JavaScript/AJAX/jQuery and Django, a high-level Python web framework that runs within GAE.

For implementing the pay-as-you-go approach in searching data we relied on Biopython [40], a rich set of Python libraries which provides the ability to deal with "things" of interest to biologists while working on the cloud. Specifically, the Entrez Programming Utilities provided by NCBI were accessed by means of the Bio.Entrez library available in Biopython. This library was made available on the cloud just making some easy changes in the source code.

## 4   BSE Functionalities

BSE is publically available at: http://biocloud-unica.appspot.com/. In what follows we will present and discuss the BSE functionality.

BSE utilizes a simple graphical user interface (GUI) that takes account for the concept of usability in presenting information. Specifically, BSE GUI is implemented by an accordion i.e. a vertically stacked list of items each of ones can be "expanded" or "stretched" to reveal the content associated with that item. There can be zero or more items expanded at a time, depending on the show/hide operation users carry on.

When a web page is loaded, the accordion expands the corresponding item into a window which contains the web page and allows users to navigate through this page. Practically, an accordion is expandable whenever needed and allows to really save some space while showing a lot of information.

By default, the first item is expanded whenever an accordion appears. Each item can be open/closed by clicking on it. A new item is added dynamically on top of the accordion to present query results.

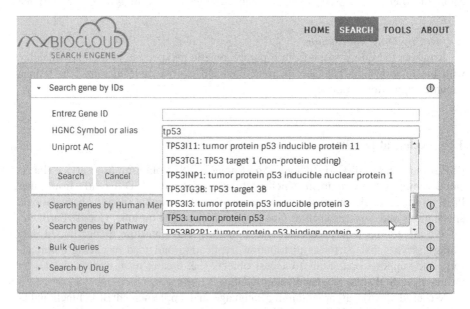

**Fig. 1.** The BSE main page with the *basic accordion*. This screenshot shows a typical search by typing the HGNC official gene symbol related to tumor protein 53 gene. Within this search context the user can also search for genes by Entrez ID, or UNIPROT accession. The Alias gene identification is supported too.

BSE implements the following accordions:

(1) The *basic accordion* is the BSE main page where each item represents a query context.
(2) The *gene accordion* is visualized whenever the user clicks on a gene identifier and allows to detail information about that gene.
(3) The *drug accordion* is visualized whenever the user clicks on the name of a drug and allows to investigate about the drug properties and effects.

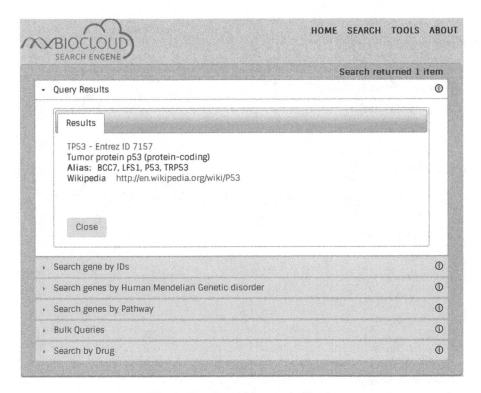

**Fig. 2.** Results to the query in Fig. 1.

In what follows we present the structure of each accordion.

Figure 1 shows the BSE main page with its *basic accordion*. The first item, corresponding to the context '*Search gene by IDs*', is expanded and shows a text field where users enter a single keyword which is the identifier of the gene they want to find. In Fig. 1 the user is typing the keyword "tp53" as standard gene identifier while BSE dynamically provides predictive suggestions by expanding the keyword "tp53" in a sliding list of its synonyms and variants. The user selects the appropriate keyword from the list, submits his query and obtains information showed in Fig. 2.

When the user clicks on the gene identifier (i.e. TP53 – Entrez ID 7157 in Fig. 2), he is redirected to the gene accordion (see Fig. 3) which details a new context to explore information about TP53. By expanding the items of this new accordion, the user can extract a series of highly detailed data and then investigate every aspect of its interest in specialized databases with a redirection that is consistent with the initial query.

For example, Fig. 4 shows the effects of expanding the item "Interaction network and Structures". Here, in order to limit the number of query results, the searching process follows a pay-as-you-go approach: the user is invited to load additional information if necessary. In this case, he interactively triggers the capture of data by clicking the "Load PDB IDs" button. Captured information is permanently stored in a buffer during 24 h and then released.

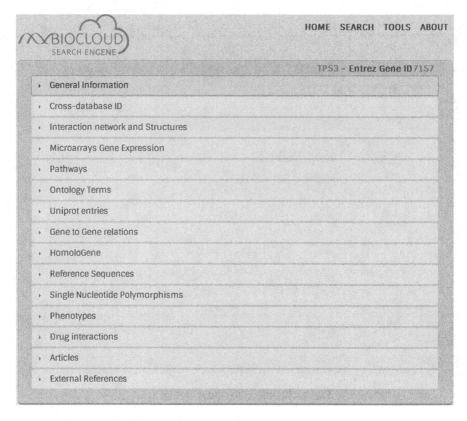

**Fig. 3.** The *gene accordion* and its items.

Figure 5 shows the results of this capture including PDB IDs, the images about related 3D structures from Protein Data bank, and FASTA Sequence of the corresponding structure. As showed on the left of Fig. 5, images can be expanded. Clicking on the sky blue arrows which have a wavy tail (see Fig. 5 on the right), the user is redirected to an external web site providing more detailed information.

The same design logic features the organization of the other search contexts of the basic accordion, i.e. *Search genes by Human Mendelian Genetic disorder, Search genes by Pathway, Bulk Queries, Search by Drug.*

In the *Search by Drug* context, as a further example, the user specifies a drug name and BSE auto-completes the user input using M.A.T.A.D.O.R. [33], a public repository which annotates relationships between human genes and drugs. Figure 6 depicts results of searching for the drug "Aspirin".

Finally, the *drug accordion* occurs whenever the user clicks on a drug name. For example, in Fig. 6, when the user clicks on "Aspirin - Pubchem ID 2244" in the window "Results", he is redirected to the drug accordion (Fig. 7) to obtain additional information.

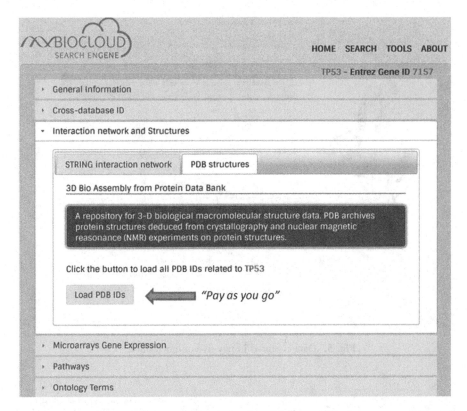

**Fig. 4.** Expansion of the item "Interaction network and Structures" in the gene accordion. The arrow shows the button to catch data in "pay as you go" fashion.

In Fig. 7, the item "General Information" is expanded and shows details about the drug "Aspirin" and the related 2D structure. The drug accordion enables searching for specific molecular information about drugs. For example, the "Protein Interactions" item shows the relationships among drugs and Gene expression products as annotated in the M.A.T.A.D.O.R. dataset [33].

The current functionalities of BSE could be extended to incorporate scalable tools for appropriate use cases in order to facilitate rapid large scale analysis of genetic information. In this direction, we are starting to implement tools which are made available easily through BSE and benefit from BSE searching capabilities.

We believe that this combination (searching-plus-tools) will allow for easy, user-friendly and transparent analysis of genetic data without requiring the user to know anything about the technical specifications of different systems (i.e. job submission, localization of web resources etc.).

Finally, the BSE user interface is unique in its focus on aggregating distributed web content in a flexible menu, a model that is highly amenable to future extension and customization by adding additional gene annotation resources, and by customizing the accordion menu to suit specific user needs.

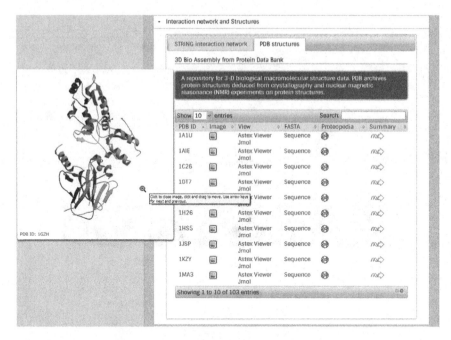

**Fig. 5.** Data captured in pay-as-you-go fashion.

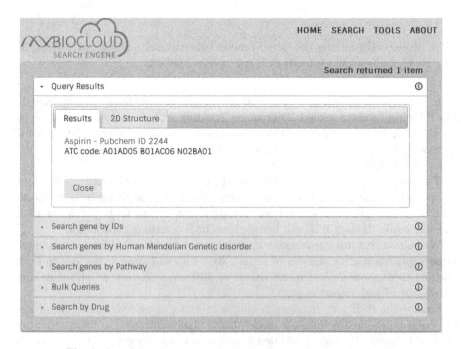

**Fig. 6.** Search by drug context in the basic accordion: query results.

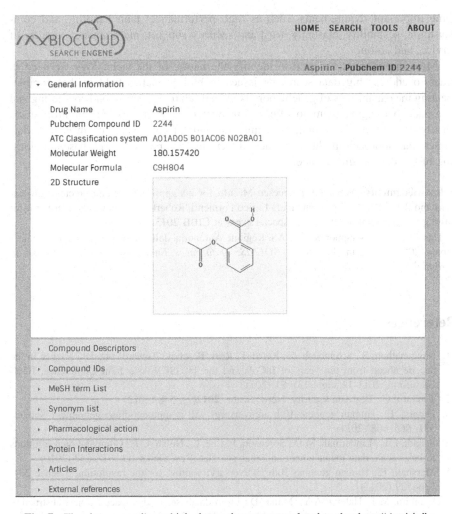

**Fig. 7.** The *drug accordion* which shows the contexts related to the drug "Aspirin".

## 5 Conclusions

Designed for people involved in the analysis of biological data (i.e. molecular biologists, biochemists, medical doctors, molecular pathologists etc.), BSE is a suitable and scalable cloud application that allows simple and advanced data searching in different databases. Going further from simply integrating content within genetic databases, as data warehousing systems do, BSE considers cloud and dataspaces the basic paradigms for effective searching big data from genomic resources. This is a unique feature of BSE.

Specifically, our work has explored how the convergence of cloud computing and dataspaces can offer both added-value service components and flexibility, making this convergence an attractive combination also for any scientific domain. BSE meets some important requirements, such as high performance, fault handling and compensation, scalability, elasticity, trust and security support, multi-tenancy, quality of service, and so on.

Most importantly, we tried to identify the nature of the technology we need in order to address big data searching issues in bioinformatics field in that complementing the capabilities of genetic portals. Albeit relatively new, cloud computing and dataspace paradigms seem to offer a prospect of new insights in bioinformatics. Finally, we are confident that, even implemented for data searching in genetic databases, our approach might reveal new directions for improving web based exploration of big data in life science.

**Acknowledgments.** We thank Francesco Masulli for his appreciations and precious suggestions and the CIBB 2013 general chairs Enrico Formenti, Roberto Tagliaferri and Ernst Wit for hosting the presentation of BSE as special event at CIBB 2013.

This research was supported by RAS, Regione Autonoma della Sardegna (Legge regionale 7 agosto 2007, n. 7), in the project *"DENIS: Dataspaces Enhancing the Next Internet in Sardinia"*.

# References

1. Ranganathan, S., Schönbach, C., Kelso, J., Rost, B., et al.: Towards big data science in the decade ahead from ten years of InCoB and the 1st ISCB-Asia Joint Conference. BMC Bioinform. 2011 **12**(suppl 13), S1 (2011)
2. Tankard, C.: Big data security. Netw. Secur. **2012**(7), 5–8 (2012)
3. Pennisi, E.: Human genome 10th anniversary: will computers crash genomics? Science **331**, 666–668 (2011)
4. Schadt, E.E., Linderman, M.D., Sorenson, J., Lee, L., Nolan, G.P.: Computational solutions to large-scale data management and analysis. Nat. Rev. Genet. **11**, 647–657 (2010)
5. Marshall, E.: Human genome 10th anniversary: waiting for the revolution. Science **331**, 526–529 (2011)
6. Franklin, M.J., Halevy, A.Y., Maier, D.: From databases to dataspaces: a new abstraction for information management. SIGMOD Rec. **34**(4), 27–33 (2005)
7. Halevy, A.Y., Franklin, M.J., Maier, D.: Principles of dataspace systems. In: Proceedings of PODS'06, pp. 1–9. ACM, New York (2006)
8. Hogue, C., Ohkawa, H., Bryant, S.: A dynamic look at structures: WWW-entrez and the molecular modeling database. Trends Biochem. Sci. **21**, 226–229 (1996)
9. Ostell, J.: The entrez search and retrieval system. The NCBI Handbook [Internet] (2002), updated 2003. http://www.ncbi.nlm.nih.gov/books/NBK21081/
10. National Center for Biotechnology Information. http://www.ncbi.nlm.nih.gov/
11. Bilofsky, H.S., Burks, C., Fickett, J.W., Goad, W.B., et al.: The GenBank genetic sequence databank. Nucl. Acids Res. **14**(1), 1–4 (1986)
12. Mizrachi, I.: GenBank: the nucleotide sequence database. The NCBI Handbook [Internet] (2002), updated 2007. http://www.ncbi.nlm.nih.gov/books/NBK21105/
13. Sayers, E.W., Barrett, T., Benson, D.A., Bolton, E., et al.: Database resources of the national center for biotechnology information. Nucl. Acids Res. **40**(Database issue), D13–D25 (2012)

14. McEntyre, J., Lipman, D.: PubMed: bridging the information gap. CMAJ **164**(9), 1317–1319 (2001)
15. Canese, K., Jentsch, J., Myers, C.: PubMed: the bibliographic database. The NCBI Handbook [Internet] (2002), updated 2003. http://www.ncbi.nlm.nih.gov/books/NBK21094/
16. Dong, X., Halevy, A.Y.: Indexing dataspaces. In: Proceedings of the 2007 ACM SIGMOD International Conference on Management of Data. SIGMOD'07, pp. 43–54. ACM, New York (2007)
17. Howe, B., Maier, D., Rayner, N., Rucker, J.: Quarrying dataspaces: schemaless profiling of unfamiliar information sources. In: Proceedings of ICDEW'08, pp. 270–277. IEEE Computer Society (2008)
18. Atzori, M., Dessì, N.: Dataspaces: where structure and schema meet. Stud. Comput. Intell. **375**, 97–119 (2011)
19. Jeffery, S.R., Franklin, M.J., Halevy, A.Y.: Pay-as-you-go user feed-back for dataspace systems. In: Proceedings of the 2008 ACM SIGMOD International Conference on Management of Data. SIGMOD'08, pp. 847–860. ACM, New York (2008)
20. Hedeler, C., Belhajjame, K., Paton, N.W., Fernandes, A.A.A., et al.: Pay-as-you-go mapping selection in dataspaces. In: Proceedings of the 2011 ACM SIGMOD International Conference on Management of Data. SIGMOD'11, pp. 1279–1282. ACM, New York (2011)
21. Chen, J., Qian, F., Yan, W., Shen, B.: Translational biomedical informatics in the cloud: present and future. BioMed. Res. Int. **2013**, 8 (2013). Article ID 658925
22. Stonebraker, M.: SQL databases v. NoSQL databases. Commun. ACM **53**(4), 10–11 (2010)
23. Sayers, E.: E-utilities quick start. Entrez Programming Utilities Help [Internet] (2008), updated 2013. http://www.ncbi.nlm.nih.gov/books/NBK25500/
24. Chambers, J., Davies, M., Gaulton, A., Hersey, A., et al.: UniChem: a unified chemical structure cross-referencing and identifier tracking system. J. Cheminform. **5**, 3 (2013)
25. The UniProt Consortium: Reorganizing the protein space at the Universal Protein Resource (UniProt). Nucl. Acids Res. **40**, D71–D75 (2012)
26. Jensen, L.J., Kuhn, M., Stark, M., Chaffron, S., et al.: STRING 8–a global view on proteins and their functional interactions in 630 organisms. Nucl. Acids Res. **37**(Database issue), D412–D416 (2009)
27. Kelder, T., Pico, A.R., Hanspers, K., van Iersel, M.P., et al.: Mining biological pathways using WikiPathways web services. PLoS ONE **4**(7), e6447 (2009)
28. Kanehisa, M., Goto, S., Sato, Y., Furumichi, M., et al.: KEGG for integration and interpretation of large-scale molecular datasets. Nucl. Acids Res. **40**, D109–D114 (2012)
29. Wu, C., MacLeod, I., Su, A.I.: BioGPS and MyGene.info: organizing online, gene-centric information. Nucl. Acids Res. **41**(Database issue), D561–D565 (2013)
30. Europe PMC. http://europepmc.org/RestfulWebService
31. NoSQL. www.nosql-database.org
32. Maglott, D., Ostell, J., Pruitt, K.D., Tatusova, T.: Entrez Gene: gene-centered information at NCBI. Nucl. Acids Res. **33**(Database issue), D54–D58 (2005)
33. Günther, S., Kuhn, M., Dunkel, M., Campillos, M., et al.: SuperTarget and Matador: resources for exploring drug-target relationships. Nucl. Acids Res. **36**(Database issue), D919–D922 (2008)
34. Croft, D., O'Kelly, G., Wu, G., Haw, R., et al.: Reactome: a database of reactions, pathways and biological processes. Nucl. Acid Res. **39**, D691–D697 (2011)
35. McKusick, V.A.: Mendelian Inheritance in Man: A Catalog of Human Genes and Genetic Disorders. Johns Hopkins University Press, Baltimore (1998)

48    N. Dessì et al.

36. Pruitt, K.D., Tatusova, T., Brown, G.R., Maglott, D.R.: NCBI Reference Sequences (RefSeq): current status, new features and genome annotation policy. Nucl. Acid Res. **40**(Database issue), D130–D135 (2012)
37. de Magalhaes, J.P.: The biology of ageing: a primer. In: Stuart-Hamilton, I. (ed.) An Introduction to Gerontology, pp. 21–47. Cambridge University Press, Cambridge (2011)
38. Yang, W., Soares, J., Greninger, P., Edelman, E.J., et al.: Genomics of Drug Sensitivity in Cancer (GDSC): a resource for therapeutic biomarker discovery in cancer cells. Nucl. Acid Res. **41**(Database issue), D955–D961 (2013)
39. Google App Engine. https://developers.google.com/appengine/
40. Biopython. www.biopython.org/

# Genomic Sequence Classification Using Probabilistic Topic Modeling

Massimo La Rosa$^{(\boxtimes)}$, Antonino Fiannaca, Riccardo Rizzo, and Alfonso Urso

ICAR-CNR, National Research Council of Italy,
via P. Castellino 111, 80131 Napoli, Italy
{larosa,fiannaca,ricrizzo,urso}@pa.icar.cnr.it

**Abstract.** Taxonomic classification of genomic sequences is usually based on evolutionary distance obtained by alignment. In this work we introduce a novel alignment-free classification approach based on probabilistic topic modeling. Using a $k$-mer (small fragments of length k) decomposition of DNA sequences and the Latent Dirichlet Allocation algorithm, we built a classifier for 16S rRNA bacterial gene sequences. We tested our method with a tenfold cross validation procedure considering a bacteria dataset of 3000 elements belonging to the most numerous bacteria phyla: Actinobacteria, Firmicutes and Proteobacteria. Experiments were carried out using complete and 400 bp long 16S sequences, in order to test the robustness of the proposed methodology. Our results, in terms of precision scores and for different number of topics, ranges from 100 %, at class level, to 77 % at genus level, for both full and 400 bp length, considering k-mers of length 8. These results demonstrate the effectiveness of the proposed approach.

**Keywords:** Genomic classification · Alignment-free analysis · 16S rRNA · DNA $k$-mers · Topic modeling · LDA

## 1 Introduction

Taxonomic classification of bacteria isolates has become of fundamental importance in biomedical and microbiological fields. The ever growing amount of biological data produced by high-throughput sequencing technologies has led to the design and implementation of bioinformatics tools in order to analyse this kind of data. Taxonomic studies of bacteria species are based on the analysis of their 16S rRNA housekeeping gene [8,24], that can be seen as a species barcode. Genomic analysis of 16S sequences for taxonomic classification purpose has been carried out in the first place by finding sequence similarities, in terms of evolutionary distance, with known species using for example alignment algorithms like BLAST [1]. More recent approaches dealt with taxonomic classification of bacteria isolates through clustering techniques, with the aim of finding a match between clusters and taxonomic categories (taxa). The clustering approach was

© Springer International Publishing Switzerland 2014
E. Formenti et al. (Eds.): CIBB 2013, LNBI 8452, pp. 49–61, 2014.
DOI: 10.1007/978-3-319-09042-9_4

done considering both evolutionary distances [17, 22] and compression–based distances [18, 19] based on the Universal Similarity Metric [23]. Compression-based approaches were also used for the study of barcode sequences [20, 21]. An exhaustive comparison of classification algorithms for 16S sequences has been done in [24]. The authors concluded that the alignment–free approach based on naive Bayesian classifier proposed by [29] and the Simrank search tool by [7] produced the best classification results of 16S rRNA gene sequences. Both approaches adopts a $k$-mer representation of genomic sequences, that uses small fragment of DNA of fixed length $k$. More in detail, the algorithm in [29] uses a $k$-mer representation of DNA sequences for training a Bayesian classifier. The resulting statistical model is then used to assign a taxonomic label to a query sequence. Simrank algorithm, on the other hand, uses the $k$-mer approach to increase the speed of the similarity searches of a query sequence against a database of known 16S sequences.

In this paper we present a novel alignment-free approach for dealing with taxonomic classification of 16S gene sequences. Our proposed methodology draws its main concepts from text mining techniques, representing a suite of algorithms to analyse set of text documents. In this work, in fact, we adopt probabilistic topic models in order to set up a 16S gene sequence classifier. A probabilistic topic model is a statistical algorithm that, according to the distribution of words in documents, is able to extract a group of recurrent meaningful themes, called *topics*, that can be used to label the documents with semantic features. Our main idea is to extract the topics from a dataset of DNA sequences and then to demonstrate that sequences sharing the same topics belong to the same taxonomic group and, consequently, it is possible to use probabilistic topic modeling in order to classify the sequences according to their taxonomic membership. The experimental results we present show high classification scores, ranging from 80 % to 100 % in terms of precision rate, at phylum, class, order and family taxonomic levels. At genus level we reached about 77 % precision score. These results were obtained considering both full length sequences (about 1200–1300 bp) and 400 bp sequences, in order to check out the robustness of our approach with respect to the sequence length. These first results represent a very encouraging outline in order to tune and improve in the near future the proposed approach in order to provide a robust and more accurate classifier for 16S sequences at all taxonomic levels.

## 2    Related Work

Probabilistic topic models are mainly used in text mining field in order to organize a corpus of documents according to a set of topics, representing the recurring themes of those documents [3, 12]. In this approach a topic represents a probability distribution over the words in the documents, so that given a topic, it is possible to have a posterior probability value that a word belonging to that topic appears in an observed document. These models are then used to train classifiers that allow, by discovering the topics of a document, to classify them.

Given an unlabeled document, we can infer if it is a law document, rather than a biological document, rather than an economic paper and so on.

Topic models have been also applied to other types of input data. For example topic models have been adapted to deal with images [2,26], audio and music [14,15], social networks [25,31]. To the best of our knowledge, the other only application of topic models to the analysis of genomic sequences has been done in [9]. There the authors used topic modeling in population genetics: they aimed at discovering a genetic signature, i.e. topics, shared by a population descending from a common ancestral parent.

## 3   Methods

In this Section we first present a brief description of probabilistic topic models, the machine learning technique we adopted in our work. Secondly we present how we adapted this technique to the analysis of 16S sequence datasets.

### 3.1   Probabilistic Topic Models

Probabilistic topic models represent a class of machine learning algorithms used in order to extract a set of meaningful themes (topics) from a corpus of documents [3,28]. Topic modeling is based on the concept of generative model for documents, which assumes a document is generated through a probabilistic procedure. Topics are considered as probability distributions over a fixed dictionary and documents can be seen as a mixture of topics. In order to create a new document, first of all a distribution over topics is fixed, then a topic is randomly chosen according to the distribution over topics, and finally a word is randomly chosen with regards to the probability distribution over the vocabulary belonging to that topic. To be more precise, topic modeling aims at discovering the best hidden structure, also known as the latent variables, that generated the corpus of documents. The latent variables are the topics, the distribution over the topics, the topic distributions over the vocabulary; the observed data are the words of the documents.

Introducing some math, we call $P(z)$ the probability distribution over topics $z$ within a document and $P(w|z)$ the probability distribution over words $w$ given the topic $z$. $P(z_i = j)$ is then the probability of sampling a word from the topic $j$ and $P(w_i|z_i = j)$ is the probability of word $w_i$ given the topic $j$. The probability of the $i$th word in a given document is then $P(w_i)$ and it is defined by means of the following conditional probability distribution over words within a document (also called posterior distribution):

$$P(w_i) = \sum_{j=1}^{T} P(w_i|z_i = j)P(z_i = j) \tag{1}$$

where T is the number of topics. Latent Dirichlet Allocation (LDA) is the simplest topic model algorithm to estimate the topics from the generative model defined in Eq. 1. Since LDA's mathematic formulation is not the main subject of this paper, please refer to the work of Blei *et al.* [4] for a full description. In this Section it is important to highlight that LDA is a probabilistic topic model algorithm that allows to infer the conditional distribution of the latent hidden variables given the observed data, i.e. the words in documents. Fitting a generative model through LDA then means to find, given a corpus of documents, the probabilities that a documents has a specific topic. The number of topics is a model parameter that has to be specified *a priori*.

### 3.2   Document Model and DNA Sequences

Probabilistic topic models, as said in Sect. 2, have been basically used to extract topics from a corpus of documents. In this Section we will explain how LDA can be used to extract meaningful information from DNA sequences. In the proposed methodology, a single DNA sequence represents a document and, consequently, a dataset of genomic sequences can be seen as a corpus of documents. In order to extract words from DNA sequences, we follow a $k$-mer decomposition. A $k$-mer is a short DNA fragment of length $k$ contained in the original sequence. By extracting all the overlapping $k$-mers, for a fixed value of $k$, from a genomic sequence, it is possible to consider the $k$-mers as the words contained in that sequence. A $k$-mer representation has been successfully adopted for the analysis of genomic sequences by many authors, like for example [5,7,10,16,27,29]. Here we consider the so called *bag-of-words* representation, i.e. we do not consider the position of the $k$-mer in the sequence. A cartoon representation of $k$-mer decomposition, for $k = 8$, is shown in Fig. 1. Given a dataset of genomic sequences, it is possible to train a probabilistic topic model, using the LDA algorithm, in order to extract the topics. Since a topic, as explained in the previous section, is a probabilistic distribution over the words in the documents, it specifies how much a document, i.e. a sequence, exhibits a certain group of words, i.e. the $k$-mers. Our main hypothesis is that similar sequences which share the same group of statistically meaningful $k$-mers, exhibits with high probability the same topic, or group of topics. Our thesis is then to demonstrate that DNA sequences sharing the same topics belong to the same taxonomic group and, consequently, it is possible to use probabilistic topic modeling in order to classify the sequences according to their taxonomic membership.

## 4   Experimental Tests

### 4.1   Bacteria Dataset

In order to validate our approach, we considered the three most populous phyla belonging to the Bacteria domain: Actinobacteria, Firmicutes and Proteobacteria. The 16S rRNA sequences were downloaded from the RDP Ribosomal Database Project II (RDP-II) [6], release 10.27. DNA sequences were selected according to the following criteria:

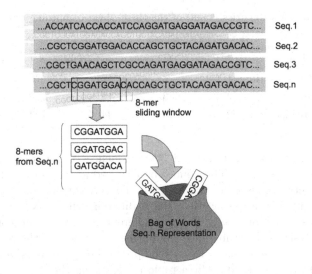

**Fig. 1.** $k$-mer decomposition (with $k = 8$) and *bag-of-words* representation for genomic sequences, adopted in our approach.

- Strain: type;
- Source: both uncultured and isolates;
- Size: $\geq 1200$ bp;
- Quality: good;
- Taxonomy: NCBI.

These options ensure sequences of high quality. In fact type strains are the best sample species; 16S average length is about 1200–1400 bp; "good" quality means the selected sequences have gone through a quality checking by RDP system. NCBI taxonomy represents the taxonomic nomenclature provided by NCBI biosystems [11]. We kept out the unclassified and "without rank" sequences. Moreover we tagged the downloaded sequences with their taxonomic category, from phylum to genus, and we considered only those taxa having at least ten elements, in order to obtain a well balanced training set. Finally, from the resulting sequences, we randomly selected 1000 sequences per phylum, so that we obtained a 16S sequences bacteria dataset of 3000 elements. In our study, we considered both full length sequences (about 1200–1400 bp) and 400 bp long sequences, obtained extracting, randomly, 400 consecutive nucleotides from the original sequences, as done in [29]. This way we want analyse he robustness of our approach with respect to the length of the genomic sequences. The taxonomic features of this dataset are summarized in Table 1.

## 4.2 Training and Testing Pipelines

Our experimental pipeline is shown in Figs. 2 and 3. In order to provide statistical significance to our experiments, we carried out a ten–fold cross validation

**Table 1.** Taxonomic categories (taxa) of the three phyla composing our 16S bacteria dataset.

| Phylum | # Class | # Order | # Family | # Genus |
|---|---|---|---|---|
| Actinobacteria | 1 | 3 | 12 | 79 |
| Firmicutes | 2 | 3 | 19 | 110 |
| Proteobacteria | 2 | 13 | 34 | 204 |

methodology. The original dataset of 3000 sequences is randomly partitioned into ten equal sized subsets. Then one subset is used as test set and the remaining nine datasets are used as training set. The cross-validation procedure is repeated ten times, considering each time a different subset as test set and the remaining ones as training set. The ten different results can be combined for statistical analysis, such for instance computing an average of the results. Looking at Fig. 2, the training set is first decomposed into its $k$-mers, and then the resulting sequences are used to fit a topic model using the LDA algorithm. In Fig. 3 it is shown the testing pipeline of our approach. Once again the test set is first decomposed through a $k$-mer approach. Then, by computing the posterior probability (see Eq. 1) of the sequences belonging to the test set, given the fitted topic model obtained during the training phase, it is possible to obtain the topics distribution over the test set. This topics distribution provides the probability that a topic is present into a test sequence. The topic distribution is one of the input data of the proposed classifier, shown in its explicit form in Fig. 4.

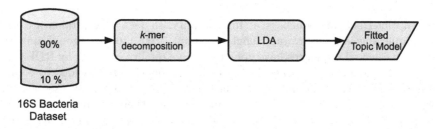

**Fig. 2.** Training pipeline. The 90 % of 16S bacteria dataset is decomposed using a $k$-mer approach and then is used as learning set of the LDA algorithm in order to fit a topic model.

### 4.3   Classification Pipeline

In order to validate our experimental tests, we adopted the following procedure, depicted in Fig. 4. Each test sequence is labeled with its most probable topic, i.e. each sequence is assigned to the topic having the highest probability value, according to the topic distribution computed during the test pipeline (Fig. 2). Then the taxonomic categories of the sequences of the training set are used to match the topics extracted by the model with the taxa. Each topic, in fact, is

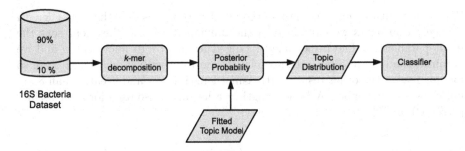

**Fig. 3.** Testing pipeline. The remaining 10 % of 16S bacteria dataset is decomposed using a $k$-mer approach and then it is computed the posterior probability of test sequences, in order to obtain their topic distributions, representing is the main input of the proposed classifier.

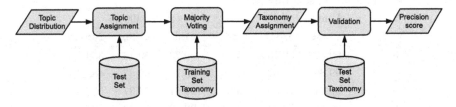

**Fig. 4.** Classification pipeline. Starting from the topic distribution, the most probable topic is assigned to each test sequence. Then, through a majority voting mechanism, the topics are matched with the taxa of original training set. Finally a precision score is computed, considering as correctly classified those test sequences whose taxon assigned to the most probable topic match with their own taxonomic category.

given a label, representing a taxon, by means of a majority voting mechanism. That means a topic is labeled with the taxon owned by the most of sequences that share that topic with the highest probability. The final validation is done considering the taxonomy of the test set and computing the precision rate of the test sequences with respect to the taxa associated to the topics. As it is well known, precision is defined as:

$$precision = \frac{\text{TP}}{\text{TP} + \text{FP}} \qquad (2)$$

where TP = true positive and FP = false positive. In our case, a sequence whose taxon match with the one of its most probable topic is considered as correctly classified (true positive); otherwise it is misclassified (false positive).

### 4.4 Results and Discussion

Our experiments have been performed using the *topicmodels* R package [13] in order to train the topic models with the LDA algorithm. We used default parameters and the Gibbs sampling method as the estimation technique for LDA.

We carried out our experiments as the number of topics and the $k$-mer length change, because, as we explained in the previous Sections, they represent the input parameters. The $k$-mer length ranged from 3 to 10 nucleotides, and we obtained the best results, in terms of precision rate, with $k$-mer length $= 8$, so for this reason the other results with $k$-mer length lesser and greater than 8 are not shown in this paper. A $k$-mer length $= 8$ has been also used for the classifier presented in [29].

**Table 2.** Classification results, in terms of precision rate and for different number of topics, for the whole 16S bacteria dataset consisting of 3000 sequences. The results are obtained with respect to `class`, `order` and `family` taxa and for full length and 400 bp sequences.

| # Topics | Precision rate (%) | | | | | |
|---|---|---|---|---|---|---|
| | Length = full | | | Length = 400 bp | | |
| | class | order | family | class | order | family |
| 3 | 99.73 | 59.13 | 35.03 | 99.73 | 59.13 | 35.03 |
| 6 | 99.80 | 70.62 | 48.50 | 99.74 | 69.87 | 48.43 |
| 9 | 99.61 | 75.80 | 55.94 | 99.61 | 74.37 | 55.79 |
| 12 | 99.41 | 78.04 | 63.14 | 99.40 | 77.28 | 62.60 |
| 15 | 99.35 | 78.87 | 68.10 | 99.35 | 78.44 | 67.94 |
| 20 | 99.79 | 83.80 | 74.21 | 99.25 | 83.23 | 73.02 |
| 25 | 99.11 | 83.78 | 77.04 | 99.78 | 83.78 | 76.84 |
| 30 | 99.43 | 87.24 | 80.18 | 99.71 | 87.85 | 79.19 |
| 70 | 100 | 88.04 | 77.95 | 100 | 85.30 | 77.51 |
| 80 | 100 | 87.35 | 76.10 | 99.60 | 85.83 | 78.28 |

In our first experiments, we used the whole bacteria dataset, composed of 3000 sequences, considering three kinds of taxa: `class`, `order` and `family`. The precision results, obtained for different number of topics, both kinds of sequence length (full and 400 bp) and averaged after a ten-fold cross validation procedure (see Sect. 4.2), are summarized in Table 2. We used a gradual growing number of topics, from 3 to 30, and then we tested our approach with a very large number of topics, i.e. 70 and 80, so that we can evaluate its performances with a small, medium and large number of topics. Looking at the `class` column, the precision values, for both sequence lengths, are always between 99 % and 100 % for each number of topics. That means the fitted topic models are able to distinguish among the five total classes of the bacteria dataset, as reported in Table 1. In particular, we reached almost a perfect score (99.73 %) even when using three topics to classify five taxa, because two of the five available classes have only one sample sequence and therefore they add only a very small amount of classification error. As for the precision rate considering the `order` category, we obtained scores lesser than 80 % when we trained the models with lesser than

**Table 3.** Classification results, in terms of precision rate and for different number of topics, for the three bacteria phyla, Actinobacteria, Firmicutes, Proteobacteria, consisting each one of 1000 sequences, considering full length sequences. The results are obtained with respect to `class`, `order`, `family` and `genus` (only for Actinobacteria) taxa.

| # Topics | Precision rate (%) - length = full | | | | | | | | | |
| --- | --- | --- | --- | --- | --- | --- | --- | --- | --- | --- |
| | Actinobacteria | | | | Firmicutes | | | Proteobacteria | | |
| | class | order | family | genus | class | order | family | class | order | family |
| 3 | 100 | 94.75 | 72.90 | 60.10 | 100 | 98.68 | 59.20 | 100 | 46.80 | 39.20 |
| 6 | 100 | 92.67 | 76.27 | 55.57 | 99.38 | 98.67 | 75.64 | 100 | 74.59 | 59.00 |
| 9 | 100 | 93.72 | 86.79 | 58.70 | 98.10 | 97.21 | 81.08 | 100 | 82.19 | 70.75 |
| 12 | 100 | 93.75 | 89.32 | 63.53 | 97.59 | 99.42 | 83.07 | 100 | 88.38 | 77.03 |
| 15 | 100 | 100 | 94.76 | 67.00 | 100 | 96.15 | 83.10 | 100 | 88.89 | 79.42 |
| 20 | 100 | 100 | 95.44 | 75.25 | 98.15 | 99.24 | 84.25 | 100 | 84.62 | 79.42 |
| 25 | 100 | 100 | 96.32 | 77.37 | 99.22 | 99.52 | 80.32 | 100 | 91.51 | 85.62 |
| 30 | 100 | 100 | 85.57 | 67.21 | 100 | 100 | 82.75 | 100 | 88.69 | 81.88 |
| 35 | 100 | 99.60 | 92.39 | 70.40 | 99.03 | 100 | 80.42 | 100 | 87.24 | 82.07 |
| 40 | 100 | 99.59 | 90.01 | 70.49 | 100 | 96.76 | 82.93 | 100 | 88.31 | 86.29 |
| 45 | 100 | 100 | 91.61 | 72.27 | 100 | 98.01 | 78.31 | 100 | 89.14 | 84.51 |
| 50 | 100 | 99.61 | 92.20 | 69.84 | 100 | 99.47 | 81.62 | 100 | 93.03 | 83.69 |
| 100 | - | - | - | 69.85 | - | - | - | - | - | - |

20 topics. In fact, we need at least 20 topics in order to correctly classify the 19 orders (Table 1) of our dataset. In this case, we obtained a precision rate between 83 % and 88 %, with a max value (88.04 %) using 70 topics. With regards to the `family` column, we obtained the highest precision score for 30 topics (80.18 %), while we expected that the best score would be for 70 and 80 topics, because we have 65 different categories to classify. In this case, with too many topics (70 and 80), the fitted models were not able to proper classify the test set and, in turn, we found better precision scores even considering a number of topics (30) lesser than the actual number of taxa (65). We obtained almost the same precision scores considering 400 bp sequences, with less than 1 % of difference with respect to the scores related to the full length sequences.

In order to further investigate the proposed approach, we retrained the topic models considering separately the three available phyla, each one consisting of 1000 sequences: Actinobacteria, Firmicutes and Proteobacteria. Once again we adopted a ten-fold cross validation procedure. The classification results, in terms of precision rate, are shown in Table 3, for full length sequences, and Table 4 for 400 bp sequences. Considering full length sequences (Table 3), precision scores for `class` and `order` taxa range from 91 % to 100 % for all three phyla, considering a number of topics at least equal to the number of categories to classify. For example, for the `order` taxon, we obtained low precision scores (lesser than 80 %) for Proteobacteria if we considered a few topics, 3 and 6, with regards to the number of different orders, i.e. 13 (Table 1). As for the `family` taxon, we reached a precision rate of about 96 % for Actinobacteria using 25 topics to classify 12 families; for Firmicutes the best precision score (84 %) is obtained using 20 topics to classify 19 families; for Proteobacteria we reachead a max precision

**Table 4.** Classification results, in terms of precision rate and for different number of topics, for the three bacteria phyla, Actinobacteria, Firmicutes, Proteobacteria, consisting each one of 1000 sequences, considering 400 bp sequences. The results are obtained with respect to class, order, family and genus (only for Actinobacteria) taxa.

| # Topics | Precision rate (%) - length = 400 bp | | | | | | | | | |
|---|---|---|---|---|---|---|---|---|---|---|
| | Actinobacteria | | | | Firmicutes | | | Proteobacteria | | |
| | class | order | family | genus | class | order | family | class | order | family |
| 3 | 100 | 93.53 | 72.60 | 59.90 | 99.19 | 96.76 | 59.10 | 99.71 | 46.60 | 38.80 |
| 6 | 100 | 91.90 | 76.04 | 54.83 | 99.47 | 95.54 | 74.50 | 100 | 72.90 | 58.50 |
| 9 | 100 | 94.06 | 86.35 | 57.19 | 99.16 | 98.15 | 81.04 | 100 | 81.38 | 70.53 |
| 12 | 100 | 94.19 | 88.36 | 59.38 | 100 | 99.43 | 82.28 | 100 | 86.27 | 74.17 |
| 15 | 100 | 100 | 91.21 | 66.97 | 99.28 | 91.03 | 80.22 | 98.28 | 87.70 | 75.36 |
| 20 | 100 | 100 | 89.85 | 71.55 | 100 | 97.94 | 81.82 | 98.72 | 87.92 | 74.90 |
| 25 | 100 | 100 | 95.47 | 74.79 | 98.11 | 98.41 | 83.82 | 100 | 84.67 | 77.21 |
| 30 | 100 | 100 | 87.10 | 68.81 | 98.63 | 93.10 | 81.21 | 100 | 87.50 | 73.96 |
| 35 | 100 | 99.59 | 91.41 | 70.24 | 99.17 | 96.08 | 78.08 | 100 | 84.53 | 75.84 |
| 40 | 100 | 99.57 | 88.28 | 71.27 | 100 | 96.73 | 80.16 | 98.25 | 84.01 | 79.63 |
| 45 | 100 | 100 | 95.57 | 72.91 | 100 | 97.04 | 77.54 | 100 | 89.45 | 76.99 |
| 50 | 100 | 99.55 | 90.64 | 70.65 | 100 | 93.91 | 80.85 | 100 | 88.84 | 79.86 |
| 100 | - | - | - | 69.35 | - | - | - | - | - | - |

score of 86 % with 40 topics and 34 families. Finally, we tested our approach considering the genus taxon only for Actinobacteria phylum. Both Firmicutes and Proteobacteria, in fact, have a large number of different genera, respectively 110 and 204, and therefore they required to train topic models with more than 100 topics. Topic models with such a number of topics are very computationally intensive. Looking at the genus column of Actinobacteria, in Table 2, the best precision scores are obtained with 25 topics. even if we expected to obtain a greater score with 100 topics because Actinobacteria phylum has 79 genera. Once again, a topic model with too many topics, in this case 100, was not able to proper classify the input dataset and, in turn, it produced an error rate greater than a topic model trained with a small number of topics with respect to the different categories to find. Moreover, topic models with hundreds of topics are very computationally expensive to train (about 24 h for a complete ten-fold cross validation procedure).

As for the results obtained with 400 bp sequences (Table 4), we obtained very similar scores with respect to the full length sequences. Precision rates, in fact, differ, at most, of less than 8 percent points and, even in this cases, we reached precision scores of about 80 %.

Our experimental results demonstrated that probabilistic topic models gives high classification results, ranging from 85 % to 100 %, when used to classify genomic sequences at class, order and family taxonomic level. These high scores are obtained using a number of topics greater than the number of actual taxonomic levels. On the other hand, with too many topics, the resulting topic models are not able to proper classify the test sets, resulting in low precision scores, about 70 %. At genus level, for Actinobacteria we scored a 77 % precision rate, but with a number of topics lesser than the number of genera. Furthermore, as we can see in Table 2, the precision results are dataset–dependent: this situation is very common when dealing with biological data, as recently investigated in [30]. Moreover precision scores exhibit little variance (less than 8 percent points) when considering 400 bp sequences, showing an high robustness of the proposed approach with respect to sequence lengths.

## 5   Conclusion and Future Work

In this paper we presented a new classification approach of 16S genomic sequences using probabilistic topic modeling. In our proposed methodology, a gene sequence represents a document and its overlapping nucleotide fragment of fixed length $k$, the $k$-mers, represent the words. We trained probabilistic topic models through LDA algorithm of 3000 16S gene sequences belonging to the most populous bacteria phyla: Actinobacteria, Firmicutes, Proteobacteria. We trained the models using both full length sequences (about 1200–1400 bp) and 400 bp long sequences. We used a ten-fold cross validation procedure and a number of topics ranging from 3 to 100, in order to analyse the classification accuracy of the fitted models with regard to the number of topics. The best classification results, in terms of precision rates, were obtained with 8-mers and with a number of topics greater than the number of different taxonomic categories of the bacteria dataset. Considering the whole bacteria dataset of 3000 elements, we obtained, for both types of sequence lengths, precision scores of 100 %, 88 % and 80 % at, respectively, class, order and family taxonomic levels. Training a separate model for each phylum, we reached more accurate results, reaching max precision scores of 100 % at class and order level for Actinobacteria and Firmicutes, and 93 % at order level for Proteobacteria. Precision rates at family level ranged from 96 % (Actinobacteria) to 86 % (Firmicutes and Proteobacteria). Finally we considered a topic model with 100 topics to classify Actinobacteria sequences at genus level, obtaining a max precision score of 77 % with a number of topics (25) lesser than the number of genera to classify (79). Considering 400 bp sequences, we obtained almost identical results, with no significative differences in the precision scores. As future work, therefore, we are going to tune and improve the proposed methodology in order to achieve higher classification results even at genus level, and then to compare our approach with other 16S classifiers, like for instance the RDP classifier [29]. Our main idea is to train a hierarchy of topic models, considering a different model at each taxonomic level (class, order, family and genus), so that the resulting classification results is given by the consensus of the single classifiers.

# References

1. Altschul, S.F., Gish, W., Miller, W., Myers, E.W., Lipman, D.J.: Basic local alignment search tool. J. Mol. Biol. **215**(3), 403–410 (1990)
2. Bart, E., Welling, M., Perona, P.: Unsupervised organization of image collections: taxonomies and beyond. IEEE Trans. Pattern Anal. Mach. Intell. **33**(11), 2302–2315 (2011)
3. Blei, D.M.: Probabilistic topic models. Commun. ACM **55**(4), 77–84 (2012)
4. Blei, D.M., Ng, A.Y., Jordan, M.I.: Latent dirichlet allocation. J. Mach. Learn. Res. **3**, 993–1022 (2003)
5. Chor, B., Horn, D., Goldman, N., Levy, Y., Massingham, T.: Genomic DNA k-mer spectra: models and modalities. Genome Biol. **10**(10), R108 (2009)
6. Cole, J.R., Wang, Q., Cardenas, E., Fish, J., Chai, B., Farris, R.J., Kulam-Syed-Mohideen, A.S., McGarrell, D.M., Marsh, T., Garrity, G.M., Tiedje, J.M.: The ribosomal database project: improved alignments and new tools for rRNA analysis. Nucleic Acids Res. **37**(Database issue), D141–D145 (2009)
7. DeSantis, T.Z., Keller, K., Karaoz, U., Alekseyenko, A.V., Singh, N.N.S., Brodie, E.L., Pei, Z., Andersen, G.L., Larsen, N.: Simrank: Rapid and sensitive general-purpose k-mer search tool. BMC Ecol. **11**, 11 (2011)
8. Drancourt, M., Berger, P., Raoult, D.: Systematic 16S rRNA gene sequencing of atypical clinical isolates identified 27 new bacterial species associated with humans. J. Clin. Microbiol. **42**(5), 2197–2202 (2004)
9. Falush, D., Stephens, M., Pritchard, J.K.: Inference of population structure using multilocus genotype data: linked loci and correlated allele frequencies. Genetics **164**(4), 1567–1587 (2003)
10. Fiannaca, A., La Rosa, M., Rizzo, R., Urso, A.: Analysis of DNA barcode sequences using neural gas and spectral representation. In: Iliadis, L., Papadopoulos, H., Jayne, C. (eds.) EANN 2013, Part II. CCIS, vol. 384, pp. 212–221. Springer, Heidelberg (2013)
11. Geer, L.Y., Marchler-Bauer, A., Geer, R.C., Han, L., He, J., He, S., Liu, C., Shi, W., Bryant, S.H.: The NCBI BioSystems database. Nucleic Acids Res. **38**(Database issue), D492–D496 (2010)
12. Griffiths, T.L., Steyvers, M.: Finding scientific topics. PNAS **101**(Suppl. 1), 5228–5235 (2004)
13. Grun, B., Hornik, K.: topicmodels: An R package for fitting topic models. J. Stat. Softw. **40**(13), 1–30 (2011)
14. Hu, D.J., Saul, L.K.: A probabilistic topic model for unsupervised learning of musical key-profiles. In: 10th International Society for Music Information Retrieval Conference (ISMIR 2009), pp. 441–446 (2009) (2009 International Society for Music Information Retrieval)
15. Kim, S., Narayanan, S., Sundaram, S.: Acoustic topic model for audio information retrieval. In: 2009 IEEE Workshop on Applications of Signal Processing to Audio and Acoustics, pp. 37–40. IEEE, October 2009
16. Kuksa, P., Pavlovic, V.: Efficient alignment-free DNA barcode analytics. BMC Bioinform. **10**(Suppl. 14), S9 (2009)
17. La Rosa, M., Di Fatta, G., Gaglio, S., Giammanco, G.M., Rizzo, R., Urso, A.M.: Soft topographic map for clustering and classification of bacteria. In: Berthold, M., Shawe-Taylor, J., Lavrač, N. (eds.) IDA 2007. LNCS, vol. 4723, pp. 332–343. Springer, Heidelberg (2007)

18. La Rosa, M., Gaglio, S., Rizzo, R., Urso, A.: Normalised compression distance and evolutionary distance of genomic sequences: comparison of clustering results. Int. J. Knowl. Eng. Soft Data Paradigms **1**(4), 345–362 (2009)

19. La Rosa, M., Rizzo, R., Urso, A.M., Gaglio, S.: Comparison of genomic sequences clustering using normalized compression distance and evolutionary distance. In: Lovrek, I., Howlett, R.J., Jain, L.C. (eds.) KES 2008, Part III. LNCS (LNAI), vol. 5179, pp. 740–746. Springer, Heidelberg (2008)

20. La Rosa, M., Fiannaca, A., Rizzo, R., Urso, A.: A study of compression–based methods for the analysis of barcode sequences. In: Peterson, L.E., Masulli, F., Russo, G. (eds.) CIBB 2012. LNCS, vol. 7845, pp. 105–116. Springer, Heidelberg (2013)

21. La Rosa, M., Fiannaca, A., Rizzo, R., Urso, A.: Alignment-free analysis of barcode sequences by means of compression-based methods. BMC Bioinform. **14**(Suppl. 7), S4 (2013)

22. La Rosa, M., Rizzo, R., Urso, A.: Soft topographic maps for clustering and classifying bacteria using housekeeping genes. Adv. Artif. Neural Syst. **2011**, 1–8 (2011)

23. Li, M., Chen, X., Li, X.: The similarity metric. IEEE Trans. Inf. Theory **50**(12), 3250–3264 (2004)

24. Liu, Z., DeSantis, T.Z., Andersen, G.L., Knight, R.: Accurate taxonomy assignments from 16S rRNA sequences produced by highly parallel pyrosequencers. Nucleic Acids Res. **36**(18), e20 (2008)

25. McCallum, A., Wang, X., Corrada-Emmanuel, A.: Topic and role discovery in social networks with experiments on enron and academic email. J. Artif. Intell. Res. **30**, 249–272 (2007)

26. Perona, P.: A Bayesian hierarchical model for learning natural scene categories. In: 2005 IEEE Computer Society Conference on Computer Vision and Pattern Recognition (CVPR'05), vol. 2, pp. 524–531. IEEE (2005)

27. Sandberg, R., Winberg, G., Bränden, C.I., Kaske, A., Ernberg, I., Cöster, J.: Capturing whole-genome characteristics in short sequences using a Naïve Bayesian classifier. Genome Res. **11**, 1404–1409 (2001)

28. Steyvers, M., Griffiths, T.: Probabilistic topic models. In: Landauer, T., McNamara, D., Dennis, S., Kintsch, W. (eds.) Handbook of Latent Semantic Analysis. Erlbaum, Hillsdale (2007)

29. Wang, Q., Garrity, G.M., Tiedje, J.M., Cole, J.R.: Naive Bayesian classifier for rapid assignment of rRNA sequences into the new bacterial taxonomy. Appl. Environ. Microbiol. **73**(16), 5261–5267 (2007)

30. Werner, J.J., Koren, O., Hugenholtz, P., DeSantis, T.Z., Walters, W.A., Caporaso, J.G., Angenent, L.T., Knight, R., Ley, R.E.: Impact of training sets on classification of high-throughput bacterial 16s rRNA gene surveys. ISME J. **6**(1), 94–103 (2012)

31. Zhou, D., Manavoglu, E., Li, J., Giles, C.L., Zha, H.: Probabilistic models for discovering e-communities. In: Proceedings of the 15th International Conference on World Wide Web - WWW '06, p. 173. ACM Press, New York (2006)

# Community Detection in Protein-Protein Interaction Networks Using Spectral and Graph Approaches

Hassan Mahmoud[1]($\boxtimes$), Francesco Masulli[1,2], Stefano Rovetta[1], and Giuseppe Russo[2]

[1] DIBRIS - Dipartimento di Informatica, Bioingegneria, Robotica e Ingegneria dei Sistemi, Università di Genova, Via Dodecaneso 35, 16146 Genova, Italy
{hassan.mahmoud,francesco.masulli,stefano.rovetta}@unige.it
[2] Sbarro Institute for Cancer Research and Molecular Medicine, College of Science and Technology, Temple University, Philadelphia, PA, USA
grusso@temple.edu

**Abstract.** Inferring significant communities of interacting proteins is a main trend of current biological research, as this task can help in revealing the functionality and the relevance of specific macromolecular assemblies or even in discovering possible proteins affecting a specific biological process. Efficient algorithms able to find suitable communities inside proteins networks may support drug discovery and diseases treatment even in earlier stages. This paper employs spectral and graph clustering methodologies for discovering protein-protein interactions communities in the *Saccharomyces cerevisiae* protein-protein interaction network.

**Keywords:** Community detection · Fuzzy C-Means clustering · Centrality · Betweenness · Spectral clustering · Modularity

## 1 Introduction

Protein-protein interactions (PPIs) occur when two or more proteins bind together in a cell, in vitro or in a living organism as the interaction interface of proteins is evolved to a specific purpose, the interactions between proteins are connected to biological functions. Not all possible PPIs occur in any cell at a given time [9]. In studies of biological networks, such as *Saccharomyces cerevisiae* PPIs [22] networks, we analyze in this paper, community detection techniques are used to extract aggregations showing dense relationships. The most used community detection algorithms can be categorized into graph based partitioning [10], hierarchical clustering [21], partitional clustering [18,24], spectral clustering [13,25], edge removal [26], and modularity based methods [27].

© Springer International Publishing Switzerland 2014
E. Formenti et al. (Eds.): CIBB 2013, LNBI 8452, pp. 62–75, 2014.
DOI: 10.1007/978-3-319-09042-9_5

It is known that proteins involved in the same cellular processes often interact with each other. Therefore, the functions of uncharacterized proteins can be predicted through comparison with the interactions of similar known proteins, and the detection of pertinent communities in PPIs networks can be used to predict the function of uncharacterized proteins based on the functions of others they are grouped with.

Adopting a suitable community detection technique in biological networks is an open problem due to various challenges of clustering approaches. Among them, there are the following: initialization criteria (e.g., choosing an initial number of clusters is required in partitional clustering like K-Means [23], while it is not needed in hierarchical clustering), accuracy (e.g., a main drawback of hierarchical clustering is the possible misclassification of some nodes [27], while removing edges may result in singleton clusters in graph bisection approach), stability (e.g., results may differ depending on the specific similarity measure used and on the random initialization of cluster centers in partitional clustering). Other challenges are complexity (e.g., deciding whether a cut exist is an NP-complete problem even for regular graphs or else for spectral clustering there is a cost concerning the computation of the first $k$ eigenvectors of their Laplacian matrix), noise sensitivity (e.g., hierarchical clustering is very sensitive to noise artifacts), and overlapping detection (e.g., the algorithm ability to detect possible overlapping between communities).

In this paper, we propose a method for communities detection based on spectral and fuzzy clustering able to infer the possible overlaps between protein communities in networks and we study its application to the analysis of the *S. cerevisiae* PPI network.

The paper is organized as follows: Graph and spectral clustering approaches for community detection are presented in Sects. 2 and 3; in Sect. 4 we propose the Fuzzy C-Means Spectral Modularity community detection method, while its application to the discovery of communities in the PPI network of *S. cerevisiae* is shown in Sect. 5; Sect. 6 contains the conclusions.

## 2    Communities Discovering in Networks Using Graph Analysis

Community detection studies [2,5,15,27] devote huge efforts to capture complex relational network structures by attempting to exploit weights between interacting nodes. In a social network weights can be a function of the relationship between linked individuals like co-authorship, duration, or friendship, while in PPI networks weights refer to the biological interactions between nodes (proteins).

A network can be represented as a weighted graph $G(V, E)$, where $V$ is the set of vertices or nodes and $E$ is the set of edges or lines and a number (weight) is assigned to each edge. The length of a path with endpoint vertices $s$ and $t$ in a graph $G(V, E)$ is the sum of the weights on its edges.

**Table 1.** The Newman's edge betweenness community detection method [26].

---

1.  Calculate the betweenness of all existing edges in the network.
2.  The edge with the highest betweenness is removed.
3.  The betweenness of all edges affected by the removal is recalculated.
4.  Repeat steps 2 and 3 are until no edges remain.

---

The art of identifying nodes having more influence over the network structure than others is referred to as a *node centrality* study. A vertex with high centrality implies that it lies on considerable fractions of shortest paths connecting vertices. Various *centrality measures* are used in network analysis such as centrality degree, closeness, betweenness, and modularity [4,14,17,20,29,31].

The *centrality degree* $C_D(v)$ of a node $v$ indicates the risk of catching the information flow and is defined as the number of links incident upon $v$, however $C_D(v)$ may be deceiving due to its locality:

$$C_D(v) = \frac{deg(v)}{n-1},$$
(1)

where $n$ is number of nodes.

*Node closeness* $C_C(v)$ is the inverse of "farness" or distance from other vertices such that $d_{G(v,t)}$ is the shortest distance between nodes $v$ and $t$ [29]:

$$C_C(v) = 1 \Big/ \sum_{t \in V} d_{G(v,t)}$$
(2)

The *betweenness* $C_B(e)$ of an edge $e$ is measured by the ratio between shortest paths linking each vertex pairs $s$ and $t$ that pass through $e$ referred as $\sigma_{st}(e)$ and all shortest paths between these pairs $\sigma_{st}$ [4,14]:

$$C_B(e) = \sum_{s,t \in V, s \neq t} \frac{\sigma_{st}(e)}{\sigma_{st}},$$
(3)

Newman [26] proposed a divisive method based on progressive removal of edges (see Table 1). Edges to be eliminated are chosen on the basis of the updated evaluation of betweenness scores after each edge removal. Betweenness can be computed for all vertices in time $O(mn)$ and requires $O(n+m)$ space for a network with $m$ edges and $n$ vertices [4]. In addition to the complexity of this approach that makes it unfeasible in application to large networks, another disadvantage is that there is no quantitative evaluation of the resultant communities.

*Network modularity* $Q$ [27] is defined as:

$$Q = \frac{1}{2m} \sum_{i,j} \left[ A_{ij} - \frac{k_i k_j}{2m} \right] \delta(c_i, c_j)$$
(4)

**Table 2.** The normalized spectral clustering method by Ng et al. [28].

---

1. **Input**: Set the initial number of clusters $k^*$, and the similarity matrix $S \in R^{n \times n}$.
2. **Processing**:
    - Compute the normalized Laplacian $L_{sym}$ given in Eq. 5.
    - Obtain the top $k$ eigenvectors $v_1, .., v_k$ of $L_{sym}$, and calculate $V \in R^{n \times k}$ by reshaping them as columns.
    - Get $U \in R^{n \times k}$ by normalizing the row sum of $V$ to 1, where $u_{ij} = v_{ij}/(\sum_k v_{ik}^2)^{\frac{1}{2}}$.
    - For $i = 1, .., n$, let $y_i \in R^k$ represents the $i^{th}$ row of $U$.
    - Apply K-Means for clustering $(y_i)_{i=1,..,n}$ points into $k$ clusters.
3. **Output**: Clusters $C_1, .., C_k$.

---

where $A_{ij}$ is the weight of edge linking vertices $i$ and $j$, $k_i = \sum_j A_{ij}$ is the degree of vertex $i$, $c_i$ is the community to which node $i$ is assigned, $m = \frac{1}{2} \sum_{ij} A_{ij}$, and $\delta(c_i, c_j)$ function is 1 if $c_i$ is the same as $c_j$ and 0 otherwise.

Network modularity is used for measuring the strength of community structure in networks and also as an objective function to maximize with suitable optimization methods. $Q$ is a scalar value ranging between $-1$ and 1. Networks with high modularity implies the existence of dense connections within communities and of sparse links between them. Although modularity suffers a resolution limit specially in case of detecting small communities, it has the advantages of not requiring prior knowledge about the number or sizes of communities, and it is capable of discovering network partitions composed of communities having different sizes.

It is worth noting that both Newman's betweenness [26] and Newman and Girvan modularity [27] approaches cannot support possible overlapping communities detection.

## 3   Communities Discovering in Networks Using Clustering Approaches

There are many approaches to clustering. Among them, the most promising for discovering communities in networks is the spectral graph partitioning, proposed by Donath and Hoffman [11].

Spectral clustering refers to methods used to cluster $n$ objects based on the evaluation of the Laplacian matrix obtained from the data similarity matrix (which is symmetric and non negative), and then in application of a clustering technique (such as K-Means) to data in a subspace spanned by the first $k^*$ eigenvectors of the Laplacian matrix. Several approaches exploit spectral theory for clustering, such as un-normalized spectral clustering by Shi and Malik [30], normalized spectral clustering by Ng et al. [28], random-walk spectral clustering by Melia and Shi [25].

It is worth to say that most of the previously mentioned spectral clustering approaches differ only in the way they calculate the Laplacian matrix $L$, and whether they apply a normalization step. One of the most popular approaches to spectral clustering is the method proposed by Ng et al. [28] (see Table 2) making use of the symmetric Laplacian matrix [6] defined as:

$$L_{sym} := D^{-\frac{1}{2}} L D^{-\frac{1}{2}} = I - D^{-\frac{1}{2}} A D^{-\frac{1}{2}}, \tag{5}$$

where $D$ is the degree matrix, and $A$ is the adjacency matrix.

When used to detect communities in networks, spectral clustering approaches, such as in Ng et al. [28] algorithm, present the following challenges:

1. The complexity of calculating eigenvectors increases with increasing number of interacting vertices.
2. The number of clusters must be specified in advance.
3. Solution are instable due to the random initialization of centroids (e.g., in case of using K-Means).
4. The method is incapable to detect possible overlaps between communities (when we use crisp clustering techniques, e.g., K-Means).

We underline that the estimation of the optimal number of communities $k^*$ is an open problem. Some intra-cluster validity indices [19] (e.g. Davies Bouldin, Dunn, etc.), or affinity measures based on eigen-gap analysis specific for spectral clustering [33] can be used for estimating the number of clusters, but they are not always reliable when used to estimate the number of communities $k^*$ inside networks.

## 4    The FSM Community Detection Method

The *Fuzzy C-Means Spectral Clustering Modularity* (or FSM) community detection method we propose in this paper applies the following three improvements to the original Ng et al. [28] spectral clustering algorithm, when used to detect communities in networks:

1. First of all, the estimation of the number of clusters is performed using the maximization of modularity procedure depicted by Neuman and Grivan in [27]; the estimated number of clusters, say $k^*$, will be used both for selecting the top eigenvectors of the Laplacian matrix, and to set the number of clusters in the clustering algorithm.
2. Then, the clustering in the affine subspace spanned by the first $k^*$ eigenvectors is obtained with the application of the Fuzzy C-Means (FCM) clustering algorithm [1,12] (see Appendix) instead of K-Means (as done in the Ng et al. [28] approach). As FCM considers that a point may belong to two or more clusters at the same time, with different membership degrees, this choice supports the detection of overlapping communities and can allow us to understand the role that each protein may play in different communities.

**Table 3.** The FSM community detection method.

---

1. Detect the number of cluster $k^*$ using modularity measurement in Eq. 4.
2. Apply the spectral clustering (e.g., Ng et al. Normalized Spectral Clustering Algorithm discussed in Sec. 3) and obtain the spectral space using top $k$ eigen vectors.
3. Cluster the resultant spectral space using Fuzzy C-Means in Eq.8.
4. Assign vertices to clusters having members larger than the threshold $\alpha$ .

---

3. After FCM, we apply an $\alpha$-cut to remove nodes with membership to discovered communities below a threshold $\alpha$. This thresholding allows us to aggregate only proteins having strong memberships, and to handle the noise and possible outliers in communities. In extreme cases it allows us to eliminate insignificant communities including nodes with low membership only. When we have an a priori knowledge on the number of possible communities, we can set the threshold $\alpha$ in order to obtain the desired number of communities. One possible criterion is to set $\alpha$ as:

$$\alpha = \frac{\eta}{l},\tag{6}$$

where $l$ is the number of expected clusters. The parameter $\eta$ (with $0 < \eta \leq 1$) is a user-selectable tuning term controlling the number of simultaneous communities to which a single node can be attributed. When $\eta = 1$, each node can belong to one community only, whereas for $\eta \to 0$ each node will be attributed to all communities. In the experiments presented in this paper we used $\eta = 0.5$.

Table 3 shows the FSM community detection method.

## 5  Saccharomyces cerevisiae PPIs Discovery

### 5.1  Dataset

In this paper we shall analyze the *S. cerevisiae*'s PPIs network. The study of the *S. cerevisiae* genetic interactions and their organization by function is the target of many bioinformatic studies [7]. *S. cerevisiae* genome sequence and a set of its deletion mutants represents about 90 % of the yeast genome. *S. cerevisiae* PPIs can be used to infer regulation of eukaryotic cells. With some 12 million base pairs and 6,466 genes, at least 31 % of *S. cerevisiae* genes have a human homologue [3].

We use the *S. cerevisiae* proteins dataset of Krogan et al. [22]. In that paper they used a tandem affinity purification to process 4,562 different tagged proteins of the yeast Saccharomyces cerevisiae. Each preparation was analyzed by

**Table 4.** Properties of four different subgraphs extracted from *S. cerevisiae* dataset [22]. For each subgraph we show the number of nodes (proteins), the number of edges and the number of estimated communities using Newman and Girvan modularity [27].

|             | SG#1 | SG#2 | SG#3 | SG#4 |
|-------------|------|------|------|------|
| Nodes       | 31   | 76   | 143  | 257  |
| Edges       | 30   | 80   | 150  | 300  |
| Communities | 2    | 5    | 12   | 19   |

both matrix-assisted laser desorption/ionization time of flight mass spectrometry and liquid chromatography tandem mass spectrometry. Then they applied an ensemble of decision trees to integrate the mass spectrometry scores and assign probabilities to the protein-protein interactions.

We can represent the dataset as an undirected weighted graph $G = (V, E)$ with $V$ vertices, corresponding to proteins, and $E$ edges indicating protein-protein interaction probabilities (weights) obtained from experiments shown in [22].

We performed our experiments on subgraphs of four different sizes obtained from *S. cerevisiae* dataset (see Table 4). The subgraphs were chosen on the basis of prior knowledge about protein involved in different biological process. For instance, protein YAL001C is the largest of six subunits of the RNA polymerase III transcription initiation factor complex (TFIIIC); part of the TauB domain of TFIIIC that binds DNA at the BoxB promoter sites of tRNA and similar genes cooperates with Tfc6p in DNA binding [16].

In this paper we report the results obtained on subgraph SG#1 only.

## 5.2   Experimental Results and Discussion

The software was developed in Matlab R2009b$^{(C)}$ under Windows 7$^{(C)}$ 32 bit. The experiments were performed on a laptop with 2.00 GHz dual-core processor and 3.25 GB of RAM.

**Application of Graph Analysis Methods.** We analyzed applied various centrality measures and algorithms on SG#1.

The evaluation of node centrality degree (Eq. 1) allows us to find the centroid YAL001C only (see Fig. 1a), while closeness (Eq. 2) discovers two central nodes: YAL007 and YDR381W (see Fig. 1b).

The results of the application of the Newman's edge betweenness community detection method [26] are shown in Fig. 2: Edges linking proteins YDR381W, YAL007C and YAL001C have highest centralities, but the application of that method is not meaningful on large subgraphs because the random null model underlying modularity becomes unreasonable.

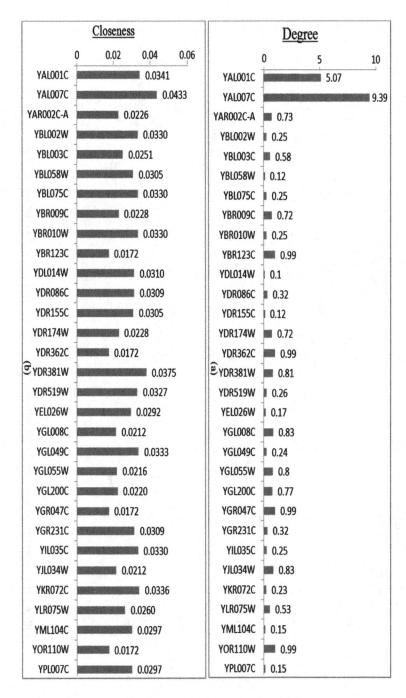

**Fig. 1.** Centrality measurements on subgraph SG#1 of the S. cerevisiae PPI network. (Left) Closeness of proteins; (right) proteins degree calculation.

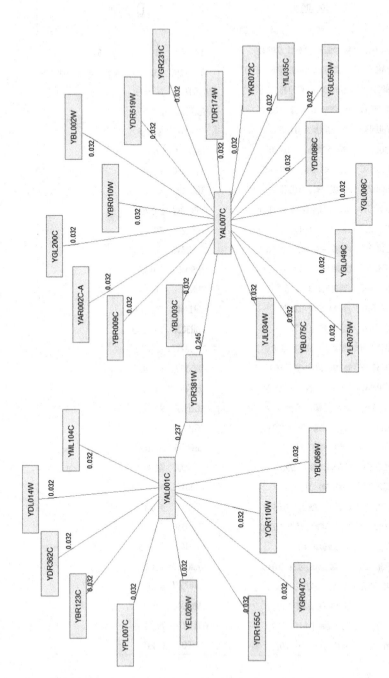

**Fig. 2.** Newman edge betweenness evaluations on subgraph SG#1 of the S.cerevisiae PPI network.

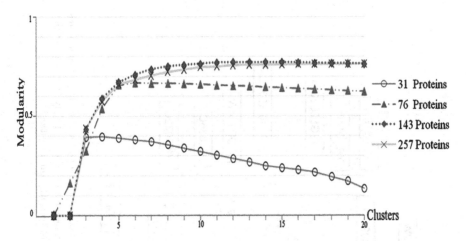

**Fig. 3.** Relationship between Newman and Girvan modularity [27] and the choice of the number of clusters (a.k.a. communities) for the four sub-graphs.

**Application of the FSM Community Detection Method.** To apply the FSM community detection method proposed in Sect. 4, we evaluated the initial number of clusters step $k^*$ as shown in Fig. 3 using Newman and Girvan modularity approach [27] on the sub-graphs on Table 4.

Figure 3 shows for each sub-graph the evaluation of modularity versus the number of clusters (or communities). The optimal values are reported on Table 4.

The we calculated the affinity or adjacency matrix $A$ between protein pairs $s$ and $t$ with entries defined as:

$$a_{s,t} = \begin{cases} 1 \text{ if } \{s,t\} \in E \\ 0 \text{ otherwise} \end{cases} \tag{7}$$

The diagonal degree matrix $D$ is obtained by calculating vertices degree. The degree $deg(v)$ of each vertex $v$ is the number of edges incident on it.

Figure 4 shows the clustering results on subgraph SG#1 obtained in spectral space using $k^* = 2$ as the number of clusters to find (corresponding to maximum modularity). The $\alpha$-cut was set to $\alpha = 0.25$.

The network is divided into two communities with centroid proteins YAL007C and YAL001C (see Fig. 4). We notice that protein YDR381W is assigned with different memberships over threshold $\alpha$ to both communities: membership $\mu(C_1) = 0.63$ to the community $C_1$ (with protein names framed with rectangles and centroid protein YAL007C), and $\mu(C_2) = 0.36$ to the community $C_2$ (with protein names framed with diamonds and centroid protein YAL001C).

We also applied the FSM community detection method to the other subgraphs of Table 4, each of them contains the preceding subgraphs. The proposed method showed robust results: for example, communities discovered from SG#2 contain those obtained from SG#1 analysis and we still discover the same overlapping proteins and fuzzy structure.

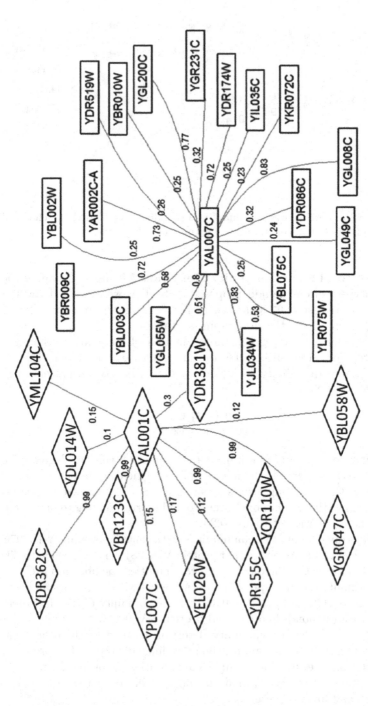

**Fig. 4.** Results of the FSM community detection method on SG#1 of 31 *S. cerevisiae* proteins edges are weighted by PPIs probabilities measured by Krogan et al. [22]. The network is partitioned into two communities, with protein labels framed, respectively, with rectangles and diamonds. Protein YDR381W is framed with an hexagon, as it has significant membership to both communities.

# 6    Conclusions

Community detection approaches allow us to extract dense aggregation of nodes from networks. They can be used in knowledge inference of different application domains such as social network analysis, and biological networks. Understanding such communities may help to understand the communication flow and the role played by entities inside networks.

This paper proposed the *Fuzzy C-Means Spectral Clustering Modularity* (or FSM) community detection method, and applies it to the identification of protein communities and detecting their possible overlapping in the *S. cerevisiae* protein-protein interaction network.

The FSM community detection method makes use of an a priori estimation of the number of communities in the network obtained by network modularity measures [27], then it applies the spectral approach [28] to project the data set in the affine space spanned by the first eigenvalues of the Laplacian of the graph associated to the network, and at that point it employs the Fuzzy C-Means algorithm [1] to cluster data and to obtain the communities, allowing possible overlapping among them.

The performed experiments on different sub-graphs of the *S. cerevisiae* protein-protein interaction network demonstrate that the proposed approach can be used with large networks, when the application of graph bases methods, such as Newman's edge betweenness community detection method [26], is not feasible, due to their high complexity.

Application of the Fuzzy C-Means algorithm in the proposed method allows us to detect possible overlapping (fuzzy) communities that cannot be identified by graph analysis methods.

## Appendix: Fuzzy C-Means Algorithm

The Fuzzy C-Means (FCM) clustering algorithm [1] performs the minimization of the following functional:

$$J_m(\mathbf{U}, Y) \equiv \sum_{i=1}^{n} \sum_{k=1}^{c} (u_{ik})^m d_{ik} \tag{8}$$

where $X = \{x_1, x_2, \ldots, x_n\}$ is a data set containing $n$ unlabeled sample points; $Y = \{y_1, y_2, \ldots, y_c\}$ is the set of the centers of clusters; $\mathbf{U} = [u_{ik}]$ is the $c \times n$ fuzzy c-partition matrix, containing the membership values of all samples to all prototypes; $m \in (1, \infty)$ is the fuzziness control parameter; $d_{ik}$ is a dissimilarity measure between data point $x_i$ and the center $y_k$ of a specific cluster $k$. Usually the Euclidean squared distance $d_{ik} \equiv \|x_i - y_k\|^2$ is employed as the dissimilarity measure.

The clustering problem can be formulated as the minimization of $J_m$ with respect to $Y$, under the normalization constraint $\sum_{k=1}^{c} u_{ik} = 1$.

The necessary conditions for minimizing $J_m$ are then:

$$y_k = \frac{\sum_{i=1}^{n}(u_{ik})^m x_i}{\sum_{i=1}^{n}(u_{ik})^m} \quad \text{for all } k, \tag{9}$$

$$u_{ik} = \left[\sum_{j=1}^{c}\left(\frac{d_{ik}}{d_{jk}}\right)^{\frac{1}{m-1}}\right]^{-1} \quad \text{for all } i, k. \tag{10}$$

The Fuzzy C-Means algorithm usually starts with a random initialization of the fuzzy c-partition matrix $\mathbf{U}$ or of the centroids $y_k$. Then, it iterates Eqs. 9 and 10 until convergence, that is usually checked by comparing the change in the position of the centroids or in the cost function with some fixed thresholds.

Note that in the limit for $m \to 1$ the fuzzy C-Means Functional $J_m$ (Eq. (8)) becomes the expectation of the K-Means global error $< E > \equiv \sum_{i=1}^{n}\sum_{k=1}^{c} u_{ik}d_{ik}$, and the FCM behaves as the classic K-Means algorithm [8,23,32].

**Acknowledgements.** Work partially funded by a grant of the University of Genova. Hassan Mahmoud is a PhD student in Computer Science at DIBRIS, University of Genova.

# References

1. Bezdek, J.C.: Pattern Recognition with Fuzzy Objective Function Algorithms. Kluwer academic publishers, Norwell (1981)
2. Bonacich, P.: Power and centrality: a family of measures. Am. J. Sociol. **92**, 1170–1182 (1987)
3. Botstein, D., Chervitz, S.A., Cherry, J.M.: Yeast as a model organism. Science **277**(5330), 1259–1260 (1997)
4. Brandes, U.: A faster algorithm for betweenness centrality. J. Math. Sociol. **25**, 163–177 (2001)
5. Brandes, U.: On variants of shortest-path betweenness centrality and their generic computation. Soc. Netw. **30**, 136–145 (2008)
6. Chung, F.: Spectral graph theory. In: Washington Conference Board of the Mathematical Sciences, pp. 849–856 (1997)
7. Costanzo, M., et al.: The genetic landscape of a cell. Science **327**(5964), 425–431 (2010)
8. Duda, R.O., Hart, P.E.: Pattern Classification and Scene Analysis. Wiley, New York (1973)
9. De Las Rivas, J., Fontanillo, C.: Protein-protein interactions essentials: key concepts to building and analyzing interactome networks. PLOS Comput. Biol. **6**(6), 1–7 (2010). doi:10.1371/journal.pcbi.1000807. e1000807
10. Ding, C. et al.: A min-max cut algorithm for graph partitioning and data clustering. In: ICDM (2001)
11. Donath, W.E., Hoffman, A.J.: Lower bounds for the partitioning of graphs. IBM J. Res. Dev. **17**(5964), 420–425 (1973)
12. Dunn, J.C.: Some recent investigations of a new fuzzy partitioning algorithm and its application to pattern classification problems. J. Cybern. **4**(2), 1–15 (1974)

13. Filippone, M., Camastra, F., Masulli, F., Rovetta, S.: A survey of kernel and spectral methods for clustering. Pattern Recogn. **41**, 176–190 (2008). ISSN: 0031–3203

14. Freeman, L.C.: A set of measures of centrality based on betweenness. Sociometry **40**, 35–41 (1977)

15. Freeman, L.C., Borgatti, S.P., White, D.R.: Centrality in valued graphs: a measure of betweenness based on network flow. Soc. Netw. **13**(2), 141–154 (1991)

16. Geiduschek, E.P., Kassavetis, G.A.: The RNA polymerase III transcription apparatus. J. Mol. Biol. **310**(1), 1–26 (2001)

17. Hage, P., Harary, F.: Eccentricity and centrality in networks. Soc. Netw. **17**, 57–63 (1995)

18. Hartigan, J.A., Wong, M.A.: Algorithm as 136: a K-means clustering algorithm. J. Roy. Stat. Soc. Ser. C Appl. Stat. **28**(1), 100–108 (1979). JSTOR 2346830

19. Jain, A.K., Dubes, R.C.: Algorithms for Clustering Data. Prentice Hall, Upper Saddle River (1988)

20. Koschützki, D., Lehmann, K.A., Peeters, L., Richter, S., Tenfelde-Podehl, D., Zlotowski, O.: Centrality indices. In: Brandes, U., Erlebach, T. (eds.) Network Analysis. LNCS, vol. 3418, pp. 16–61. Springer, Heidelberg (2005)

21. Krause, A., et al.: Large scale hierarchical clustering of protein sequences. BMC Bioinf. **6**, 6–15 (2005)

22. Krogan, N., et al.: Global landscape of protein complexes in the yeast Saccharomyces cerevisiae. Nature **440**, 637–643 (2006)

23. Lloyd, S.P.: Least square quantization in PCM, Bell telephone laboratories, Murray Hill (1957). Reprinted. In: IEEE Trans. Inf. Theor. **28**(2), 129–137 (1982)

24. Mahmoud, H., Masulli, F., Rovetta, S.: Feature-based medical image registration using a fuzzy clustering segmentation approach. In: Peterson, L.E., Masulli, F., Russo, G. (eds.) CIBB 2012. LNCS, vol. 7845, pp. 37–47. Springer, Heidelberg (2013)

25. Meila, M., Shi, J.: A random walks view of spectral segmentation. In: Artificial Intelligence and Statistics AISTATS (2001)

26. Newman, M.E.J.: Detecting community structure in networks. Eur. Phys. J. B: Condens. Matter **38**, 321–330 (2004)

27. Newman, M.E.J., Girvan, M.: Finding and evaluating community structure in networks. Phys. Rev. E **69**(2), 026113 (2004)

28. Ng, J., Jordan, M.I., Weiss, Y.: On spectral clustering: analysis and an algorithm. In: Proceedings of Neural Information Processing Systems, pp. 849–856 (2002)

29. Sabidussi, G.: The centrality index of a graph. Psychometrika **31**, 581–603 (1966)

30. Shi, J., Malik, J.: Normalized cuts and image segmentation. IEEE Trans. Pattern Anal. Mach. Intell. **22**, 888–905 (2000)

31. Shimbel, A.: Structural parameters of communication networks. Bull. Math. Biophys. **15**, 501–507 (1953)

32. Steinhaus, H.: Sur la division des corp materiels en parties. Bull. Acad. Polon. Sci **1**, 801–804 (1956)

33. Von Luxburg, U.: A tutorial on spectral clustering. Stat. Comput. **17**, 395–416 (2007)

# Weighting Scheme Methods
# for Enhanced Genomic Annotation Prediction

Pietro Pinoli[(✉)], Davide Chicco, and Marco Masseroli

Dipartimento di Elettronica, Informazione e Bioingegneria,
Politecnico di Milano, Piazza Leonardo Da Vinci 32, 20133 Milan, Italy
{pinoli,davide.chicco,masseroli}elet.polimi.it

**Abstract.** Functional genomic annotation data banks, which store the associations between genes (or a gene products) and terms of controlled vocabularies describing their features, are paramount in computational biology. Despite their undeniable importance, these data sources cannot be considered neither complete nor totally accurate; in their curated updates often new annotations are added and some of their annotations are revised. In this scenario, computational methods that are able to quicken the curation process of such data banks are very important. To this end, the Latent Semantic Indexing (LSI) by Singular Value Decomposition, and its Semantically IMproved (SIM) variant, have shown to be able to predict novel functional annotations from a set of available ones. In this work, we propose a further improvement of those techniques, based on a preparatory weighting of the associations between genes (or a gene products) and functional annotation terms. We tested the effectiveness of our approach on nine Gene Ontology annotation datasets. The results demonstrated that this technique is able to improve novel annotation predictions.

**Keywords:** Biomolecular annotation prediction · Weighting schemes · Latent Semantic Indexing · Singular Value Decomposition · Semantic analysis

## 1 Introduction

Thanks to the new technologies introduced in recent years, we have been witnessing an exponential growth in biomedical and biomolecular information, with a large amount of data becoming available for investigation. Among them, those that describe the current biomedical knowledge in a controlled and computable form are the most valuable. In particular the associations of a gene (or a gene product) to one or more controlled vocabulary terms, which describe its functional properties, are paramount to perform in silico analyses and being able to biologically interpret experimental results. Some consortia maintain lists of controlled terms and the sets of genes and proteins annotated to them. We refer to those associations as *functional annotations*, which are stored in annotation databases. When, as often occurs, semantic relations between annotation terms

© Springer International Publishing Switzerland 2014
E. Formenti et al. (Eds.): CIBB 2013, LNBI 8452, pp. 76–89, 2014.
DOI: 10.1007/978-3-319-09042-9_6

are also available (i.e. terms are organized in an ontological structure), only the annotations of biomolecular entities to the most specific terms describing their properties are stored in such databases. Their less specific annotations are implicitly defined and can be automatically derived from the ontology structure.

With the progression of the biomedical knowledge, new annotations are continuously added or revised, and annotation databases are usually neither complete or totally accurate. In this scenario, tools that are able to improve the quality of available annotations, both in coverage and accuracy, are very useful. In recent years, scientists have proposed several of such tools [1]. In 2007, Drachici et al. [2] proposed the Latent Semantic Indexing (LSI) algorithm, enhanced by means of data weighting schemes, to predict novel human gene annotations. In 2011, we proposed the Semantic IMprovement (SIM) algorithm [3], which is an improved version of LSI based on clustering and semantic similarity of genes. Here, we present an extension of these methods, featuring significant improvements through the use of novel weighting schemes.

After this introduction the paper is organized as follows. In Sect. 2 we describe both the data warehouse containing the annotation data that we considered and the *annotation unfolding* method, which was used to derive all the ontological gene annotations available from their most specific ones stored in the data warehouse. In Sect. 3, first we present the novel weighting schemes that we defined and their application to an annotation matrix. Then, we briefly describe the algorithms that we considered for annotation prediction (Latent Semantic Indexing and Semantically Improved Latent Semantic Indexing), applied on a weighted annotation matrix. In Sect. 4 we describe the procedure that we used to test correctness and reliability of our predictions, and in Sect. 5 we report and discuss some of the obtained results. Finally, in Sect. 6 we draw some conclusions about the presented work and outline some future development.

## 2 Annotation Data Considered

In order to have easy access to subsequent versions of gene annotations to be used as input to the considered algorithms or to evaluate the results that the algorithms provided, we took advantage of the Genomic and Proteomic Data Warehouse (GPDW) [4,5].

### 2.1 Genomic and Proteomic Data Warehouse

In the GPDW several controlled terminologies and ontologies, which describe genes and gene products related biomolecular process, functionalities and phenotypes, are stored together with their numerous annotations of genes and proteins of many organisms imported from several well known biomolecular databases. A software framework manages creation and bimonthly updating of the GPDW with new data releases that become available in the original databases; subsequent GPDW versions are stored for comparison. In addition, the GPDW framework provides mechanisms for automatically checking the quality of the

**Table 1.** Quantitative characteristics of the considered annotation datasets

| | Gallus gallus | | | Bos taurus | | | Danio rerio | | |
|---|---|---|---|---|---|---|---|---|---|
| | CC | MF | BP | CC | MF | BP | CC | MF | BP |
| July 2009 | | | | | | | | | |
| # genes | 260 | 308 | 273 | 497 | 541 | 511 | 430 | 692 | 1,525 |
| # features | 147 | 223 | 526 | 234 | 420 | 929 | 137 | 306 | 958 |
| # annotations | 477 | 505 | 735 | 921 | 931 | 1555 | 601 | 919 | 4,080 |
| May 2013 | | | | | | | | | |
| # genes | 775 | 690 | 898 | 1,675 | 1,419 | 1,685 | 1,460 | 1,674 | 3,342 |
| # features | 257 | 363 | 1,891 | 423 | 571 | 3,050 | 330 | 553 | 2,794 |
| # annotations | 1,560 | 1,399 | 3,917 | 3,405 | 2,781 | 7,031 | 2,432 | 2,814 | 11,008 |
| $\Delta$ annotations between GPDW versions | | | | | | | | | |
| #$\Delta$ annotations | 1,083 | 894 | 3,182 | 2,484 | 1,850 | 5,476 | 1,831 | 1,895 | 6,928 |
| %$\Delta$ annotations | 227.04 | 177.03 | 432.92 | 269.70 | 198.71 | 52.15 | 304.66 | 206.20 | 169.80 |

imported data (both in accuracy and consistency). These quality procedures are able to reduce, if not eliminate, errors and inconsistencies often present in annotation data; thus, they provide more clean data to our methods, which can then learn more consistent and coherent prediction models.

In the performed tests, we run our predictions on gene annotations from the July 2009 version of the GPDW, and compared the obtained predictions to the same annotations available in the May 2013 version of the GPDW. We focused on the Gene Ontology (GO) [6] annotations of *Gallus gallus (red junglefowl)*, *Bos taurus (cattle)* and *Danio rerio (zebrafish)* genes. We chose these organisms because their gene annotations to the three GO ontologies (Biological Process, Molecular Function, Cellular Component) included a representative number of annotations and involved genes and terms (i.e. gene features). Actually, we did not use all the annotations present in the GPDW, but for each organism we filtered out the less reliable ones (according to the GO annotation *evidence*) and the ones related to very rare annotation terms (i.e. associated with less than three genes). We denote this filtering phase as *term pruning*. Table 1 provides a quantitative description of the annotation sets that we used.

## 2.2  Annotation Unfolding

Curators of ontological annotations always use the most specific ontology terms to describe a given feature of an annotated gene or gene product. Such annotations, named direct annotations, are the only ones available in biomolecular databases. Yet, according to the ontology structure, when a gene or gene product is annotated to an ontological term, it is implicitly annotated also to all the ancestor terms of the directly annotated term. Such annotations to the ancestor terms, which describe more generic features, are named indirect annotations.

In order to be able to provide all direct and indirect annotations, when ontological annotations are imported in the GPDW from the original databases, the ontological relationships between the annotation terms are also imported and the ontology structure is computed and stored in the GPDW. Thus, at run time, when the GPDW is queried for all the ontological annotations of a biomolecular entity, the direct annotations stored in the GPDW for that biomolecular entity are unfolded according to the ontology structure, the related indirect annotations are inferred and all direct and indirect annotations are made available. We refer to this process as *annotation unfolding* [7].

## 3   Computational Methods

In this Section we describe the techniques that we considered to predict novel gene annotations starting from the available ones. In Subsect. 3.1 we describe the novel weighting schemes that we propose and how we apply them to an annotation matrix. In Subsects. 3.2 and 3.3 we briefly describe the prediction methods (thoroughly illustrated in [3]) that we considered and propose to enhance with the new weighting schemes.

### 3.1   Weighting Schemes for Annotation Matrices

Let $G$ be a set of genes (or gene products) and $F$ a set of their features described by terms of the controlled vocabulary of an ontology. We can then build the binary annotation matrix $\mathbf{A} \in \{0,1\}^{|G| \times |F|}$, whose rows correspond to genes and columns correspond to feature terms. An element $\mathbf{A}(g,f)$ of this matrix can assume value 1, if the gene $g$ is annotated to the term $f$ or to one of its descendants in the considered ontology structure; otherwise it assumes value 0. Previous works in information retrieval [8] have shown that it is possible to improve the performance of a predictive system in terms of precision by using a real representation, instead of a binary one, of the considered data. In our case such representation can be obtained by weighting the associations between genes and feature terms. Towards this goal, different weighting schemes can be used. The intuitions behind these schemes are the following:

- if a feature term is included in multiple unfolding paths to the ontology root from any of the ontology terms directly associated with a given gene, then this feature is more strongly related to that gene than a feature represented by a term of the same ontology that is included in less unfolding paths;
- if a feature is associated only with a limited number of genes, it is a good discriminator of the genes; thus, the annotation to the term representing that feature should be considered more important, because it brings a higher amount of information.

These two concepts can be expressed by using two statistics: *term-frequency* (*tf*) and *inverse-gene-frequency* (*igf*). *Term-frequency* measures how important a feature (represented by an ontology term) is to a certain gene; it provides a local weight. For each gene $g$ and feature $f$, $tf(g,f) = 1 + N$, where $N$ is the number of child-terms of $f$ which are associated with the gene $g$, both directly or indirectly (i.e. added by the unfolding procedure).

*Inverse-gene-frequency* measures how important an annotation to a certain feature is; it provides a global weight that decreases the relevance of common features, while increasing the relevance of the rare ones. For each feature $f$ we can compute:

$$igf(f) = \ln \frac{|G|}{|genes\ annotated\ to f|} \tag{1}$$

These two statistics can be combined in order to build different weighting schemes. Draghici et al. [2] proposed some of these schemes. Here we introduce new schemes that differ from the previously proposed ones because of the way in which the above frequencies are computed. In keeping with Draghici and colleagues [2], we refer to each of our weighting schemes with a three letter code: the first letter specifies the local weight used by the scheme (e.g. *tf*), the second letter denotes the used global weight (e.g. *igf*), and the last letter indicates which normalization function is applied (Table 2).

Each weighting schema can be applied to the annotation matrix by (a) multiplying every gene annotation by the corresponding local weight factor and by the global weight factor of the specific annotation term and (b) normalizing the new obtained real valued annotations by the normalization factor. For example the schema named NTM uses as local schema the *tf* directly, *igf* as global weighting and the maximum normalization. That is, for each entry $\mathbf{A}(g, f)$ of the annotation matrix, the correspondent entry in the weighted annotation matrix $\mathbf{W}(g, f)$ is computed as:

$$\mathbf{W}(g, f) = tf(g, f) \cdot igf(f) \cdot \frac{1}{max_{f'}\{tf(g, f') \cdot igf(f')\}} \tag{2}$$

The possible combinations of the weight and normalization functions described in Table 2 lead to 9 weighting schemes, but only 7 of them are distinct (notice that the NTM and NTC schemes are equal to the MTM and MTC schemes respectively).

**Table 2.** Weighting schemes

| Code | Name | Description |
|------|------|-------------|
| *Local Weight* | | |
| N | No-Transformation | $\forall f, g : w_{loc} = tf(g, f)$ |
| M | Maximum | $\forall f, g : w_{loc} = tf(g, f) / max_{f'} tf(g, f')$ |
| A | Augmented | $\forall f, g : w_{loc} = 0.5 + 0.5 \cdot (tf(g, f)/max_{f'} tf(g, f'))$ |
| *Global Weight* | | |
| T | Term Weight | $\forall f : w_{glob} = igf(f)$ |
| *Normalization* | | |
| N | None | Normalization factor is not used |
| M | Maximum | $w_{norm}(g, f) = w(g, f) / max_{f'} w(g, f')$ |
| C | Cosine | $w_{norm}(g, f) = w(g, f) / \sqrt{\sum_{f'} w(g, f')}$ |

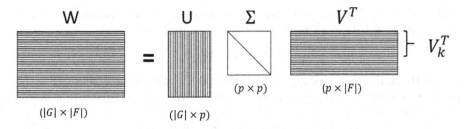

**Fig. 1.** The Singular Value Decomposition of a weighted annotation matrix $\mathbf{W}$.

## 3.2 Latent Semantic Indexing by Singular Value Decomposition

Let $\mathbf{W}$ be a weighted annotation matrix. The Singular Value Decomposition (SVD) of such matrix is $\mathbf{W} = \mathbf{U}\mathbf{\Sigma}\mathbf{V}^T$, where $\mathbf{V}$ is an orthonormal $|G| \times p$ real matrix whose columns are the left-singular vectors of $\mathbf{W}$, $\mathbf{\Sigma} = diag(\sigma_1, \sigma_2, \ldots, \sigma_p)$, with $0 \leq \sigma_p \leq \sigma_{p-1} \leq \ldots \leq \sigma_1$, is a $p \times p$ diagonal matrix of the non-negative singular values of $\mathbf{W}$, $\mathbf{V}$ is an orthonormal $|F| \times p$ real matrix whose columns are the right-singular vectors of $\mathbf{W}$, and $p = min(|G|, |F|)$. Figure 1 shows an example of SVD decomposition.

Given the annotation profile of a gene (i.e. a vector $\mathbf{a} \in \{0, 1\}^{|F|}$ of all the gene annotations to the terms of the considered vocabulary), the $\mathbf{V}_k$ matrix, composed of the first $k < p$ columns of the $\mathbf{V}$ matrix, can be used to obtain the predicted annotation profile $\hat{\mathbf{a}} \in \mathbb{R}^{|F|}$ for the gene. It can be computed as follows:

$$\hat{\mathbf{a}}^T = \mathbf{a}^T \mathbf{V}_k \mathbf{V}_k^T \tag{3}$$

Then, every real valued $\hat{\mathbf{a}}_{ij}$ annotation obtained can be compared to the corresponding $\mathbf{a}_{ij}$ annotation in the matrix $\mathbf{A}$. If, given a threshold $\tau \in [0, 1]$, $\hat{\mathbf{a}}_{ij} > \tau$ (i.e. the annotation of the gene $i$ to the term $j$ is suggested by the computational method used) and $\mathbf{a}_{ij} = 0$ (i.e. that annotation is not given as input to the computational method), such annotation is defined as a *predicted annotation*. The threshold $\tau$ can be chosen on the basis of how many most likely annotations are desired to be obtained. In our tests, we chose $\tau = 0.5$.

## 3.3 Semantic IMprovement (SIM)

An improvement of the Latent Semantic Indexing (LSI) method has been proposed with the SIM method [9], which clusters the genes and considers their semantic similarity. The LSI method implicitly uses a global term-to-term correlation matrix $\mathbf{T} = \mathbf{W}^T\mathbf{W}$, which is estimated from the whole corpus of available annotations (notice that, according to the SVD, $\mathbf{V}$ is composed of the eigenvectors of $\mathbf{T}$). Instead, authors in [9] proposed an adaptive approach, which clusters genes based on their original annotation profiles and for every cluster estimates a distinct correlation matrix $\mathbf{T}_c$, $c = 0, 1, \ldots, C$, where $C$ denotes the number of clusters and $\mathbf{T}_0 = \mathbf{T}$. For each matrix $\mathbf{T}_c$, a set of $k$ eigenvectors $\hat{\mathbf{V}}_{c,k}$ is computed. Then, for each gene annotation profile $\mathbf{a}$, the $\mathbf{V}_{c,k}$ matrices are used to compute $C + 1$ different predicted annotation profiles:

$$\hat{\mathbf{a}}_c^T = \mathbf{a}^T \mathbf{V}_{c,k} \mathbf{V}_{c,k}^T \tag{4}$$

Among them, the best predicted annotation profile of the gene is selected as the one that minimizes the $L2$-norm with respect to the original annotation profile of the gene:

$$\hat{\mathbf{a}} = \arg \min_{c=0,\dots,C} \|\hat{\mathbf{a}}_c - \mathbf{a}\|_2 \tag{5}$$

In order to build the correlation matrices $\mathbf{T}_c$, the genes can be clustered based on their functional similarity, expressed by their functional annotations. To this end, we can exploit the SVD of the $\mathbf{W}$ matrix, as suggested in [10]. In fact, each column $\mathbf{u}_c$ of the matrix $\mathbf{U}$ represents a cluster and the value $\mathbf{U}(i, c)$ represents the degree of membership of the gene $i$ to the $c^{th}$ cluster (notice how every gene can belong to more than one cluster at the same time, with different degrees of membership). The estimation of $\mathbf{T}_c$ proceeds as follows. First, a modified gene-to-term matrix $\mathbf{W}_c = \mathbf{C}_c \mathbf{W}$ is generated, where $\mathbf{C}_c \in \mathbb{R}^{|G| \times |G|}$ is a diagonal matrix with the entries of $\mathbf{u}_c$ along the main diagonal. Then, $\mathbf{T}_c = \mathbf{W}_c^T \mathbf{W}_c$ is computed. A more accurate clustering can be obtained by incorporating the functional similarity between the annotation terms.

As for LSI, once the output $\hat{\mathbf{A}}$ matrix is obtained, every $\mathbf{a}_{ij}$ element of the input annotation $\mathbf{A}$ matrix is compared with its corresponding $\hat{\mathbf{a}}_{ij}$ element in the output matrix, and the *predicted annotations* are found. Figure 2 provides an overview of the SIM method with weighting schema.

### 3.4   Anomaly Correction

According to the hierarchical structure of an ontology and the *True Path Rule* of annotations [11], when a gene (or gene products) is annotated to an ontological term, it must be also annotated to all its ancestor terms. Yet, this is not guaranteed to hold for the predicted annotation profiles computed by the (weighted) LSI or SIM methods. In fact, since they do not consider the ontology structure and the predicted annotations are defined based on the comparison between the predicted annotation values and a threshold, these methods might suggest that a gene shall be annotated to a term, but not to one or more of its ancestor terms.

To avoid such anomalous cases and hence to violate the *True Path Rule*, we defined an *anomaly correction* procedure. It is based on the transformation of the predicted annotation values, so that the value of an annotation between a gene and an ontological term can not be greater than any of the annotation values of the gene to any of the term's ancestors. Our procedure consists in updating parent-term annotation values with the greater annotation value of their child-terms. An iterative process is carried on from annotation ontology leaf nodes to the ontology root node, in order to correct possible annotation values that could create anomalies for some threshold values. Compared to more complex *anomaly correction* methods previously proposed [12], this procedure can be efficiently applied also on large annotation datasets.

1. Retrieve the gene annotations and populate the $\mathbf{A}(g, f)$ annotation matrix
2. Compute the $tf(g, f)$ and $igf(f)$ statistics on $\mathbf{A}(g, f)$
3. Define the weighting schema to be used and calculate its weighting factors based on $tf(g, f)$ and $igf(f)$
4. Compute the $\mathbf{W}(g, f)$ weighted annotation matrix by multiplying every element of $\mathbf{A}(g, f)$ by its corresponding weighting factor
5. Define a number $C \in \{1, 2, 3, \ldots N\}$ of clusters and a truncation level $k \in \mathbb{N}$
6. Consider only the first $\mathbf{U}_c$ columns of the $\mathbf{U}$ matrix, where $c \in \{1, ..., C\}$ and $\mathbf{W} = \mathbf{U\Sigma V}^T$
7. Use the membership degree $\mathbf{U}(\text{i,c})$ of gene $i$ to cluster $c$, to cluster the genes
8. For each cluster $c$:
   (a) Generate $\mathbf{W}_c = \mathbf{C}_c\mathbf{W}$, where $\mathbf{C}_c = \text{diag}(\mathbf{u}_c)$ and $\mathbf{u}_c$ is the $c^{th}$ column of the $\mathbf{U}$ matrix
   (b) Compute the correlation matrix, $\mathbf{T}_c = \mathbf{W}_c{}^T\mathbf{W}_c$
   (c) Compute the set of $k$ eigenvectors $\mathbf{V}_{c,k}$ of $\mathbf{T}_c$
9. For each annotation profile $\mathbf{a}$ of a gene $g$:
   (a) For each cluster $c$, compute the vector:

$$\hat{\mathbf{a}_c}{}^T = \mathbf{a}^T\mathbf{V}_{c,k}\mathbf{V}_{c,k}^T$$

   (b) Among the $C$ predicted annotation profiles $\hat{\mathbf{a}_c}$ of gene $g$, chose as best prediction the one that minimizes the *L-2 norm*:

$$\hat{\mathbf{a}} = \arg \min_{c=0,\ldots,C} \|\mathbf{a} - \mathbf{a_c}\|_2$$

10. Consider as predicted annotations of gene $g$ to the feature terms $f$ the ones that have $\mathbf{a}(f) = 0$ and $\hat{\mathbf{a}}(f) > \tau$, with $\tau \in [0, 1]$ range

**Fig. 2.** Overview of the SIM algorithm with weighting schema

## 3.5 Computational Complexity

The time complexity of LSI and SIM methods is strictly related to the complexity of the Singular Value Decomposition. According to [13], the computational complexity of the Lanczos-based approach, which we have been using to approximate the actual SVD of the matrix $W$, is $O(k \times (nnz + |F|))$, where $nnz$ is the number of non-zero elements of $W$, $k$ is the SVD truncation level introduced in Subsect. 3.2 and $|F|$ is the number of features.

The overall complexity of the weighing scheme is $O(|G| \times |F| \times U)$, where $U$ is the complexity of the unfolding operation, which we use so as to estimate the number of ancestor terms. This operation depends on the number of annotations and on the ontological DAG depth; in the worst case it can be estimated to be $O(|F|^2)$.

Given the usually high dimension of the considered annotation matrix, the computational time could be a critical aspect of a prediction algorithm; so we tried to optimize as much as possible the software implementation of the algorithms, in particular by exploiting data parallelism where possible.

## 4    Validation of Annotation Predictions

We used the algorithms described above, with and without weighting schemes, to analyze sets of gene functional annotations and predict new gene annotations based on the available ones. In order to assess the actual improvement given by the weighting schemes, we compared the predictions given by the weighted versions of the LSI and SIM algorithms with the ones given by their unweighted versions. To this end, we evaluated the quality of a set of predicted annotations as follows:

1. we extracted the annotations to be used as input to the prediction algorithm from an outdated version of the GPDW (July 2009 version in our tests)
2. we excluded from those annotations the less reliable ones, i.e. the annotations with IEA (Inferred from Electronic Annotation) *evidence* code
3. by running the prediction algorithm, we got a list of predicted annotations ordered by their confidence value (i.e. their corresponding $\hat{\mathbf{A}}(g, f)$ value)
4. we toke into account the top $P$ predictions (we used $P = 250$) and we counted how many predictions among these $P$ were found confirmed in the updated version of the GPDW (May 2013 version in our tests).

We depict the validation procedure workflow in Fig. 3. Notice that selecting the top $P$ predictions corresponds to set, for each dataset, the value of the threshold $\tau$ that leads to the selection of $P$ predictions. We preferred to fix the number of predictions rather than the value of $\tau$, so as to homogeneously compare the results given by different weighting schemes.

## 5    Validation Results

Test results for LSI and SIM methods are reported in Tables 3 and 4, respectively. The obtained validation results show that the proposed weighting schemes generally improve annotation predictions in the tested datasets, especially in the smallest ones: in the *Gallus gallus* datasets, the weighted variants of LSI outperform the unweighted one in 14 out of 21 cases (66.7 %) and the weighted variants of SIM outperform the classical one in 16 out of 21 cases (76.2 %). Overall, not all the schemes provide equal benefits; in particular, the schemes with the *No-Transformation* local weight seem to be the best ones. Conversely, the schemes with the *Augmented* local weight seem to provide a poorer improvement.

Although the obtained validation results showed that the weighting schemes are able to improve the classical predictive methods, they can be considered only an underestimation of the real number of corrected predictions; in facts, the computed list of predictions could include several additional biologically correct annotations that have not been included yet in the annotation databases, and hence in the GPDW, or which have not been even discovered yet.

As an example of our predictions, we report in Fig. 4 the Directed Acyclic Graph (DAG) of the Gene Ontology Cellular Component terms predicted by the LSI method, with the NTN weighting schema, as associated with the *Heat*

**Table 3.** Validation results of the predictions obtained by the LSI algorithm, without or with different weighting schemes: amount of the top 250 predicted gene annotations to the GO Biological Process (BP), Molecular Function (MF) and Cellular Component (CC) terms in the *Gallus gallus*, *Bos taurus* and *Danio rerio* datasets that have been found confirmed in the updated GPDW version. In **bold** the cases where a weighting schema outperforms the unweighted method. k: SVD truncation level.

| LSI, k=50 | | | | | | | | |
| --- | --- | --- | --- | --- | --- | --- | --- | --- |
| Dataset | None | NTN | NTM | NTC | MTN | ATN | ATC | ATM |
| *Gallus gallus* - BP | 49 | 43 | **63** | **55** | **63** | 44 | 41 | **53** |
| *Gallus gallus* - MF | 20 | 12 | **24** | 18 | 18 | 9 | **31** | **27** |
| *Gallus gallus* - CC | 24 | **48** | **31** | **35** | **43** | **45** | **39** | **33** |
| *Bos taurus* - BP | 61 | 53 | 56 | 34 | 39 | 53 | 34 | 47 |
| *Bos taurus* - MF | 36 | 14 | 33 | **37** | 31 | 20 | **38** | 35 |
| *Bos taurus* - CC | 75 | **102** | **81** | **79** | **78** | 70 | 53 | **87** |
| *Danio rerio* - BP | 36 | **44** | 36 | 36 | **47** | **51** | **42** | **60** |
| *Danio rerio* - MF | 43 | 26 | 37 | 34 | 25 | **48** | **45** | 36 |
| *Danio rerio* - CC | 40 | **45** | **42** | **41** | 33 | **45** | 27 | 28 |
| Total | 384 | **387** | **403** | 372 | 377 | **385** | 350 | **406** |

**Table 4.** Validation results of the predictions obtained by the SIM algorithm, without or with different weighting schemes: amount of the top 250 predicted gene annotations to the GO Biological Process (BP), Molecular Function (MF) and Cellular Component (CC) terms in the *Gallus gallus*, *Bos taurus* and *Danio rerio* datasets that have been found confirmed in the updated GPDW version. In **bold** the cases where a weighting schema outperforms the unweighted method. k: SVD truncation level; C: number of gene clusters.

| SIM, k=50, C=3 | | | | | | | | |
| --- | --- | --- | --- | --- | --- | --- | --- | --- |
| Dataset | None | NTN | NTM | NTC | MTN | ATN | ATC | ATM |
| *Gallus gallus* - BP | 45 | **86** | **56** | **50** | **62** | 24 | 18 | 43 |
| *Gallus gallus* - MF | 5 | **24** | **21** | **26** | **33** | **6** | **25** | **15** |
| *Gallus gallus* - CC | 31 | **50** | **32** | 28 | **42** | 30 | **42** | **38** |
| *Bos taurus* - BP | 40 | **55** | **49** | **64** | **53** | 35 | 27 | 22 |
| *Bos taurus* - MF | 39 | 28 | 28 | 28 | 20 | **41** | 32 | 31 |
| *Bos taurus* - CC | 76 | **91** | **89** | 70 | 53 | 22 | 70 | 63 |
| *Danio rerio* - BP | 55 | 35 | **70** | **83** | **56** | 36 | 36 | **63** |
| *Danio rerio* - MF | 32 | **35** | **58** | 22 | 27 | **48** | 28 | 21 |
| *Danio rerio* - CC | 33 | **44** | **41** | **38** | 22 | 26 | 24 | 24 |
| Total | 356 | **447** | **444** | **409** | **378** | 268 | 302 | 320 |

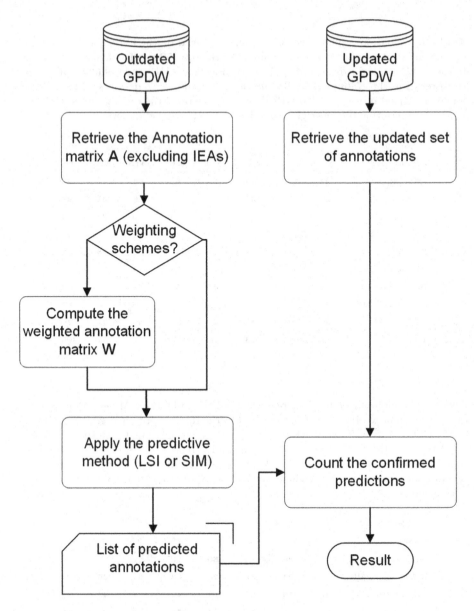

**Fig. 3.** Workflow of the validation process that we implemented in order to assess the quality of a set of predictions.

*shock 70kDa protein 8 (HSPA8)* gene (Entrez Gene ID: *395853*) of the *Gallus gallus* organism. One may notice that for this gene the predictions comprise new annotations to GO terms that constitute a DAG independent branch, including the *organelle (GO:0043226)* and *membrane-bounded organelle (GO:0043227)* terms, and new annotations to five GO terms that constitute a DAG sub-branch.

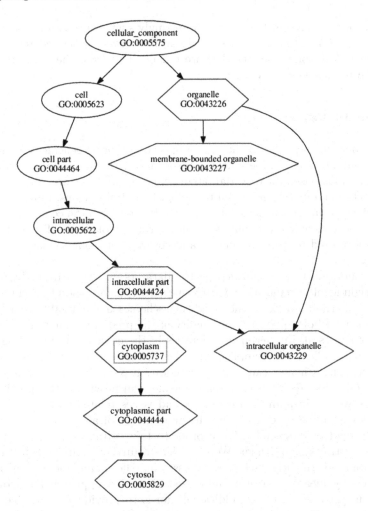

**Fig. 4.** Direct Acyclic Graph of the GO Cellular Component annotation terms predicted by the LSI method, weighted with the NTN schema, as associated with the *Heat shock 70kDa protein 8 (HSPA8)* gene (Entrez Gene ID: *395853*) of the *Gallus gallus* organism. The black circle elements represent the annotation terms of the *HSPA8* gene that were already present in the considered dataset (from the GPDW version of July 2009), while the blue hexagon elements represent those terms of the predicted annotations. The *intracellular part (GO:0044424)* and *cytoplasm (GO:005737)* predicted annotation terms, in the green rectangles, were found and confirmed in the validation dataset (from the updated GPDW version of May 2013). The other predicted annotation terms might be added in the GPDW database in the future.

Out of these seven predicted annotations for the *Gallus gallus (HSPA8)* gene, we found two of them in the validation dataset (from the May 2013 updated version of the GPDW database), i.e. the annotations to *intracellular part (GO:0044424)* and *cytoplasm (GO:005737)*.

The shown prediction DAG very interestingly suggests the annotation of the *Gallus gallus* (*HSPA8*) gene to several features never associated to this gene before, including some features that are independent from the ones previously annotated to this gene.

# 6   Conclusions

In this paper, we have shown that the proposed weighting schemes enhance the predicted annotations. These schemes, associated with classical annotation prediction methods based on vectorial latent semantic analysis, are able to predict novel functional annotations useful to biologists and physicians. In addition, the proposed weighting schemes are also independent on both the considered organism and the controlled vocabulary used for the annotations. Thus, they can be profitably used to improve quality and coverage of biomolecular annotation databases.

The validation tests performed on the datasets extracted from the GPDW show a significant enhancement with respect to the correspondent unweighted prediction methods, making the weighting schemes a remarkable tool for the improvement of both quantity and quality of the predicted annotations. Nevertheless, our tests also underlined difficulties in the choice of the best weighting schemes, particularly when applied on wider datasets. It seems that mostly proposed weighting schemes can improve annotation prediction results on smaller datasets. On wider datasets only some schemes can provide better predictions.

Future work will address advantages and issues related to annotation prediction, for example by considering the annotations of all three GO ontologies jointly, instead of independently, in order to take advantage of potential correlations existing among them. We also plan to further verify the effectiveness of the proposed weighting schemes by assessing the quality of the top ranked predictions by means of a literature-based validation approach (as shown for example in [14]). At last, our additional aim is to provide Web service access to our implemented method and integrate such Web service with other available services within the Search Computing framework [15], in order to provide support for answering complex life science questions [16,17].

**Acknowledgments.** This research is part of the Search Computing project (2008–2013) funded by the European Research Council (ERC), IDEAS Advanced Grant. The authors would like to thank Luke Lloyd-Jones for the help in the revision of the English style of the text.

# References

1. Pandey, G., Kumar, V., Steinbach, M.: Computational approaches for protein function prediction: a survey. Technical report, Department of Computer Science and Engineering, University of Minnesota, Minneapolis, MN, USA (2006)

2. Draghici S., Done B., Purvesh K., Done A.: Semantic analysis of genome annotations using weighting schemes. In: Proceedings of IEEE Symposium on Computational Intelligence in Bioinformatics and Computational Biology, pp. 212–218 (2007)

3. Chicco, D., Tagliasacchi, M., Masseroli, M.: Biomolecular annotation prediction through information integration, In: Proceedings of CIBB 2011 - Computational Intelligence Methods for Bioinformatics and Biostatistics, pp. 1–8 (2011)

4. Canakoglu, A., Ghisalberti, G., Masseroli, M.: Integration of biomolecular interaction data in a genomic and proteomic data warehouse to support biomedical knowledge discovery. In: Biganzoli, E., Vellido, A., Ambrogi, F., Tagliaferri, R. (eds.) CIBB 2011. LNCS, vol. 7548, pp. 112–126. Springer, Heidelberg (2012)

5. Pessina, F., Masseroli, M., Canakoglu, A.: Visual composition of complex queries on an integrative genomic and proteomic data warehouse. Engineering 5(10B), 94–98 (2013)

6. Gene Ontology Consortium: Creating the gene ontology resource: design and implementation. Genome Res. 11, 1425–1433 (2001)

7. Masseroli, M., Tagliasacchi, M.: Web resources for gene list analysis in biomedicine. In: Lazakidou, A. (ed.) Web-based Applications in Health Care and Biomedicine. Annals of Information Systems Series, vol. 7, pp. 117–141. Springer, Heidelberg (2010)

8. Salton, G.: Introduction to Modern Information Retrieval. McGraw-Hill, New York (1983)

9. Masseroli, M., Tagliasacchi, M., Chicco, D.: Semantically improved genome-wide prediction of gene ontology annotations. In: Proceedings of the 11th IEEE International Conference on Intelligent Systems Design and Applications, pp. 1080–1085 (2011)

10. Drineas, P.: Clustering large graphs via the singular values decomposition: theoretical advances in data clustering. Mach. Learn. 56, 9–33 (2004). (guest editors: Nina Mishra and Rajeev Motwani)

11. Tanoue, J., Yoshikawa, M., Uemura, S.: The GeneAround GO viewer. Bioinformatics 18, 1705–1706 (2002)

12. Masseroli, M., Tagliasacchi, M.: Anomaly-free prediction of gene ontology annotations using Bayesian networks. In: 9th IEEE International Conference on Bioinformatics and Bioengineering, pp. 107–114 (2009)

13. Chen, J., Saad, Y.: Lanczos vector versus singular vectors for effective dimension reduction. Technical report, Department of Computer Science and Engineering, University of Minnesota, Minneapolis, MN, USA (2008)

14. Nuzzo, A., Mulas, F., Gabetta, M., Arbustini, E., Zupan, B., Larizza, C., Bellazzi, R.: Text Mining approaches for automated literature knowledge extraction and representation. Stud. Health Technol. Inf. 160(Pt 2), 954–958 (2010)

15. Ceri, S.: Chapter 1: search computing. In: Ceri, S., Brambilla, M. (eds.) Search Computing. LNCS, vol. 5950, pp. 3–10. Springer, Heidelberg (2010)

16. Chicco, D.: Integration of bioinformatics web services through the search computing technology. Technical report, Dipartimento di Elettronica e Informazione, Politecnico di Milano, Milan, Italy (2012)

17. Masseroli, M., Ghisalberti, G., Ceri, S.: Bio-search computing: integration and global ranking of bioinformatics search results. J. Integr. Bioinf. 8(166), 1–9 (2011)

# French Flag Tracking by Morphogenetic Simulation Under Developmental Constraints

Abdoulaye Sarr$^{(\boxtimes)}$, Alexandra Fronville, Pascal Ballet, and Vincent Rodin

UMR CNRS 6285, Lab-STICC, CID, IHSEV, Computer Science Department,
Université de Brest, 20 avenue Le Gorgeu, 29200 Brest, France
abdoulaye.sarr@univ-brest.fr

**Abstract.** Below the influence of the mechanical cues and genetic expression, constraints underlying the developmental process play a key role in forms' emergence. Theses constraints lead to cells' differentiation and sometimes determine the directions of cells growth. To better understand these phenomena, we present in this paper our work focused primarily on a development of a mathematical model. A one which takes into account the *co-evolution* of cellular dynamics with it's environment. To study the influence of the developmental constraints, we have developed algorithms to make and explore a base of genomes. The purpose of this exploration is first to check conditions under which specific genes are activated. Then, this exploration allows us to follow the conditions of emergence of some patterns that lead to a specific shape. From our model, we found a genome that can generate the *French flag*. With this *French flag* pattern and its genome starting, we addressed the following question: is there another genome in the simulated base that achieves the same shape, i.e. the *French flag* pattern?

**Keywords:** Mathematical modelling · Simulation of biological systems · Morphogenesis · Multi-agent system · French flag problem

## 1 Introduction

### 1.1 Morphogenesis: Emerging of Interests

Biomedical science has undergone a remarkable evolution during this last decade. Advances and innovations in biotechnology, more particularly in microscopy and imaging, have provided a large amount of data. And these data include all levels of the biological organization. In 2007, Melani and al. achieved a tracking of cell's nuclei and the identification of cell's divisions in live zebrafish's embryos. They used 3D+time images acquired by confocal laser scanning microscopy [13]. These kind of data allowed new description in details of many components and structures of living organisms. Observations noticed from these data like geometrical segmentations during cells' proliferation have raised relevant issues in mathematical and numerical point of view. But experimental complexity usually restricts observations to a single or very restricted spatial or temporal scales.

© Springer International Publishing Switzerland 2014
E. Formenti et al. (Eds.): CIBB 2013, LNBI 8452, pp. 90–106, 2014.
DOI: 10.1007/978-3-319-09042-9_7

Thus, one of the purposes of mathematical and computational models in biology is to reconstruct integrated models from this large amount of data gathered at different scales. So that, the dynamical interactions between different levels of the biological organization is taken into account. In cellular proliferation models, it means to consider the cell as a place of integration of causalities and downgrades.

For this purpose, *tensegrity model* considers biomechanical forces between cells and the extracellular matrix. The stretching of cells adhering to the extracellular matrix may result from local reshuffle in this latter. According to this model, growth-generated strains and pressures in developing tissues regulate morphogenesis throughout development [9]. It is therefore the biomechanical forces which play a key role in this model of morphogenesis. For example by modulating cell's differentiation, influencing the direction of division or deforming tissues. However, the question of cells' diversity even arises before the acquisition of shape [16]. Indeed, when the embryo has only a few pairs of cells, we can already see a diversification of biochemical content or a diversification of embryonic cells' morphology. That may be the result of genetic and molecular interactions. Indeed, the emergence of shapes also stems from the acquisition of differential properties, from cells' mobility and genes' expression throughout their development.

Moreover, *Artificial Regulatory Networks* also allow modelling morphogenesis [15]. They define a series of regulatory genes and structural genes. The first consist of a network of rules determining the evolution of the system. And the latter are intended to each generate a simple specific pattern. They can be seen as a dynamical system following different trajectories in a state space [10]. However, even if the detailed knowledge of genomic sequences allows to determine where and when different genes are expressed in the embryo, it is insufficient to understand how the organism emerge [14].

### 1.2   Below Genetic Expression

So, we have looked forward to learn more about the emergent properties of cellular organization. And especially, the importance of cellular dynamics in the emergence and evolution of shapes, both in mathematical and numerical point of view.

A Multicellular organism is a complex system which can be defined as a composition of a significant number of elements interacting locally to produce a global behaviour. Complex systems are also characterized by a high capacity of self-adaptation and self-organization similarly to multicellular organisms. They can evolve and learn through feedbacks between their external environment and their internal architecture. And according to Doursat [5], whether inanimate structures or living organisms, all processes of shape's emergence are instances of decentralized morphological self-organization. When cells evolve, they modify their organism which in its turn impacts their behaviour. This is what biologists mean by *co-evolution*. Epigenetic considers that this coupling between organism and environment can not be ignored in understanding the development of living organisms [19].

In mathematics, the *viability theory* [2] offers concepts and methods to control a dynamical system in a given fixed environment, in order to maintain it in a set of constraints of viability. Applied to morphogenesis, this means that we should have at least one co-viable evolution of the cells' state and their environment based on each state-environment pair. This formalization allows us to establish feedback rules in terms of changes in growth direction and cells' differentiation. This mathematical model was formalized in [7]. So, in our computational model, every cell has its own rules monitored by a set of controls and stays aware of its neighbourhood. This ensures cells to be autonomous while being aware of a wrong evolution of their dynamic, and consequently, to know when to implement feedback mechanisms. The integration of the behaviours of each component of the system allows to determine its global state in a finite time.

In this paper, we are going first to establish the mathematical formalization of the model described above (Sect. 2) before presenting our morphogenesis simulation tool (Sect. 3). Then, we aim to better understand the influences of shape's emergence on cellular dynamics. To do so, we have developed here original algorithms based on the mathematical model. The main goal of these algorithms is to explore and simulate a base formed by all possible genomes from a same set of genes. Genome and gene don't have the same meaning here as in Biology. In our view, a gene carries a set of properties for the cell, such as colour, reproductive age, energies' minimum levels, viscosity, rigidity, maximum allowable pressure, tolerated stress threshold, choice of dividing direction etc. In a given simulation, it's allowable to not define some of theses properties. Genes are designated by their index. A genome is defined as a suit of genes' indexes. During a simulation, when a gene in a genome is activated, cells adopt the properties defined by this gene. The exploration of the base of genomes allows us to verify the conditions under which certain genes are activated. And more specifically, if these conditions are unique for a targeted pattern (Sect. 4). We test the algorithms on our morphogenesis simulation tool with the *French flag* as being the form to reach (Sect. 5). Finally, we conclude before highlighting some relevant applications and future prospects we could give to them (Sect. 6).

## 2    Morphological Dynamic of Cells

### 2.1    Mathematical Model

We proposed in [7] a mathematical model of morphogenesis through a formalization of cellular dynamics in the context of mutational and morphological analysis [1,12]. It provides an extension of differential equations in a metric space instead of the classical Euclidean space $\mathbb{R}^N$. At the tissue level, we have a large group of connected cells, of a same cellular type, performing a specific function [18]. Therefore, the behaviour of cellular tissue can be seen as a result of a bottom-up process of cellular dynamic. A minor change on tissue implies that cells have implemented dynamics where each cell can not only "move" but also can multiply, die or stay quiescent (see Fig. 1).

**Fig. 1.** Multivalued analysis to formalize a cell that multiplies and moves

$M \subset \mathbb{R}^3$ denotes the cells' containment and represents the complement of the *vitellus*[1].

$K \subset \mathbb{R}^3$ representing tissue cells, the cells are denoted by $x \in K \subset \mathbb{R}^3$.

If we restrict morphogenesis in the plan,

$$\mathcal{D} := \{(1,0), (-1.0)(0,1), (0,-1)\}$$

For convenience, we note:

$$\mathcal{D} := \{1, 3, 2, 4\}$$

denotes the set of 4 planes directions and

$$\overline{\mathcal{D}} := \mathcal{D} \cup \{(0,0)\} \cup \emptyset$$

means the 6 "extended" directions.

For morphogenesis in the space $\mathbb{R}^3$,

$$\mathcal{D} := \{(1,0,0), (-1,0,0), (0,1,0), (0,-1,0), (0,0,1), (0,0,-1)\}$$

For convenience:

$$\mathcal{D} := \{1, 3, 2, 4, 5, 6\}$$

denotes the set of six directions and $\overline{\mathcal{D}} := \mathcal{D} \cup \{(0,0,0)\} \cup \emptyset$ means the eight "extended" directions.

We note

$$\Xi_M(K, x) := \{u \in \mathcal{D}, \text{ such that } x + u \in \{x\} \cup (M \setminus K)\}$$

and

$$R_M(K, x) := \Xi_M(K, x) \times \Xi_M(K, x).$$

Then we introduce the correspondence

$$\Psi(x, u, v) := \{x + u\} \cup \{x + v\}_{(u,v) \in R_M(K,x)}.$$

The morphological dynamic $\Phi_M$ is then defined by

$$\Phi_M(K) := \bigcup_{x \in K} \bigcup_{(u,v) \in R_M(K,x)} \Psi(x, u, v) \tag{1}$$

And the discrete morphological dynamic $K_{n+1} = \Phi_M(K_n)$.

---

[1] In biology, the vitellus is the energy reserve used by the embryo during its development.

This gives the different cases of cell behaviour:

1. *apoptosis*, obtained by taking $(\emptyset, \emptyset) \in R_M(K, x)$ since $\Psi(x, \emptyset, \emptyset) = \emptyset \cup \emptyset = \emptyset$
2. *migration* by taking $u \in \mathcal{D}$ and $v = \emptyset$ or $u = \emptyset$ and $v \in \mathcal{D}$ or further $u = v$
3. *stationarity*, which is a migration obtained by taking $u$ and $v$ equal to $(0, 0, 0)$
4. *cell division* by taking $u := (0, 0, 0)$ and $v \in \Xi_M(K, x)$ (or vice-versa)
5. *division and migration* by taking $u \in \Xi_M(K, x)$ and $v \in \Xi_M(K, x)$

We introduce an equivalence relation on the directions

$$u \equiv_x v \text{ if and only if } x + u = x + v$$

which we denote by $\mu$ and $\nu$ the representatives, noting that by construction, for every pair $(\mu, \nu)$ the equivalence class, for all $u \in \mu$ and $v \in \nu$, $\Psi(x, \mu, \nu) = \Psi(x, u, v)$ does not depend on the choice of directions belonging to equivalence classes.

Because two cells can not occupy the same position, at most they just select one extensive direction in each class.

The correspondence of regulation is defined by the quotient set:

$$\Theta_M(K, x) := R_M(K, x)/\equiv_x \tag{2}$$

The morphological dynamics $\Phi_M$ is always defined by

$$
\begin{aligned}
\Phi_M(K) &:= \bigcup_{x \in K} \bigcup_{(\mu, \nu)) \in \Theta_M(K, x)} \Psi(x, \mu, \nu) \\
&= \bigcup_{x \in K} \bigcup_{(u, v) \in R_M(K, x)} \Psi(x, u, v)
\end{aligned}
\tag{3}
$$

In the case of a discrete dynamic, it is defined by the sequences of control $(u_n, v_n)$ associated to $K_n$ to be able to define $K_{(n+1)}$.

Implementing this formalization is to set the viable directions throughout the developmental process whatever underlying constraints.

## 2.2   Shapes Emergence

Thanks to our mathematical model, here is a code achieving the three first cellular segmentations:
$\forall x \in K_1 = \{(0, 0, 0)\}$, the first route choice for step 1 (see Fig. 2) may be:

$$U(1, x) = U(1) = [1, 3, 2, 4, 5, 6, 0]$$

$\forall x \in K_2 = \{(0, 0, 0), (1, 0, 0)\}$, the second route choice for step 2 (see Fig. 3) may be:

$$U(2, x) = U(2) = [2, 4, 1, 3, 5, 6, 0]$$

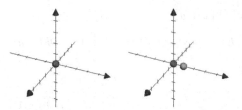

**Fig. 2.** $U(1, x) = U(1) = [1, 3, 2, 4, 5, 6, 0]$ means that the first axis of segmentation is $x$-$axis$ and the direction is right

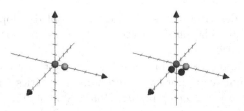

**Fig. 3.** $U(2, x) = U(2) = [2, 4, 1, 3, 5, 6, 0]$ means that the second axis of segmentation is $y$-$axis$ and the direction is forward

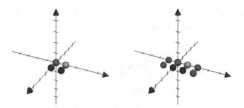

**Fig. 4.** $U(3, x) = [1, 3, 2, 4, 5, 6, 0]$ means that the third axis of segmentation is $x$-$axis$ and the direction is right in the first choice and left in second one

$\forall x \in K_3 = \{(0, 0, 0), (1, 0, 0), (0, 1, 0), (1, 1, 0)\}$, the third route choice for step 3 (see Fig. 4) may be:

$$U(3, x) = U(3) = U(1) = [1, 3, 2, 4, 5, 6, 0]$$

For the following steps, the route choices can be chosen among any possible ones. Considering, for instance, the step $n$, a route choice $v$ is the one which enable segmentation of any cell in $K$. That means the route choice which let the system evolving:

$$u \in \Xi_M(K_n, x) = \{u | x + u \in M \setminus K_n\}$$

The list of directions in the route choice $U(n, x)$ for which, the place is empty in $(M \setminus K_n)$ is $R(n, x)$.

Then for each $x \in K_n$, we have a direction $u \in R(n, x)$.

And $\psi(x, 0, u) = \{x\} \cup \{x + u\}$

$$K_{n+1} = \Phi_M(K_n) = \bigcup_{\substack{x \in K_n \\ u \in R(n,x)}} \Psi(x, u)$$

$$= \bigcup_{x \in K_n} \Psi(x, R(n, x))$$

## 3    Simulation Tool

In this section, we describe `Dyncell` which is a tool developed for simulating generative systems [6]. The platform was created to experiment our theoretical model of morphogenesis.

The program is implemented in `C++` using a tool kit of Virtual Reality: `ARéVi` [4,17]. `ARéVi` is a simulation library of autonomous entities with a `3D` rendering developed at *European Center of Virtual Reality* (Brest, France).

### 3.1    Architecture

The architecture of our simulation tool is based on the concept of shape. All classes inherit from a generic parent class called `Form` (see Fig. 5). During simulation, each instantiated object is a form by definition.

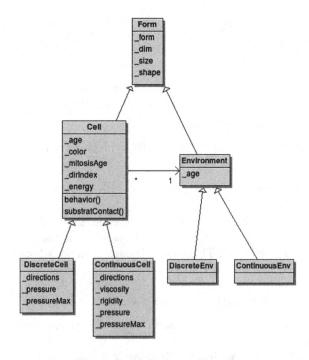

**Fig. 5.** Dyncell classes diagram

This class has two main subclasses: `Environment` and `Cell`, which are the two types of objects to be instantiated before starting any simulation.

Running a simulation always consists of instantiating an object `Environment` and some `Cells` objects evolving in this environment.

Class `Cell` is responsible for representing cells. Then, an object `Cell` is defined by its *age, color, division speed* and *energy amount.* During mitosis, another `Cell` object is created next to the mother `Cell` object that triggered the division. The position of the daughter `Cell` depends on the route choice taken by its mother, as defined in the mathematical model.

Environment is a `Form` object which has a lifetime. It is used to delimit the space in which cells' population operates or to exert external forces on them.

## 3.2   Features

The order of scheduling has a significant impact on the results of the simulation [3,11]. It determines how local interactions (self-organization) have been held. Thus, how and when the system reaches the final shape. Different behaviours can be observed in virtual models depending on the scheduling mode. When modelling natural systems, asynchronism seems more appropriate. In morphogenesis, for instance, synchrony is likely to cause false correlations between cells. In `ARéVi`, scheduling of agents is handled implicitly in asynchronous and stochastic way. Routines of each agent are executed sequentially and entirely, the same number of times. However the running order of agents is randomly determined. We set as a basic principle that cells are autonomous agents and ignorant of the whole system even if they can perceive their environment and adapt accordingly. They are represented on screen by spheres and can proliferate in a discrete environment (cellular automaton) or in a "continuous" one. In the latter case, the movement of cells is more precisely described and we can have more complex interactions.

A graphical interface has been implemented in order to set the options of the simulation. It allows dynamic change of parameters and selecting mechanisms (e.g. apoptosis) that will be active/inactive during the simulation.

Options are available to allow choice between 2D/3D, discrete or continuous simulations. The size and shape of both the environment and the cells can also be defined and adjusted.

Spatial constraints are crucial for evolution of the cells. If they are too heavy, the cell is not viable as it can no longer divide. A maximal constraint parameter sets up a threshold below which the cell can remain viable.

To account for the influence of the environment, a parameter is defined as the maximum number of cells that a cell is able to push when it divides. When the current strain of the cell is greater than the maximum stress threshold, the cell can no longer divide. To stay alive, a cell can undertake two modes of mitosis. Firstly, the cell chooses to divide in the direction where the stress is less intense. Secondly, it takes a predetermined direction.

It is also possible to assign an amount of energy to each `Cell`. The basic idea is to consider that an organism accumulates a certain reserve of energy by consuming its environment's foods. A percentage of this reserve is used to maintain its structure and growth. The remaining part is used for its maturation and reproduction. Thus, in `Dyncell`, at each timestep, cells lose an energy amount intended for their structure maintaining. Moreover, during mitosis, the energy level is shared between the mother and the daughter `Cell`. A `Cell` dies when its energy level becomes too low.

## 4   Algorithms

At the initial stage of living organisms' creation, all cells have the same genome. But during the developmental process, they don't all express the same genes. Why has the dynamic of these cells changed to lead to their differentiation? What are the factors that come into play? What is the mechanism by which this takes place?

These are the questions that underlie any understanding of collective self-organization and self-adaptation mechanisms that target then achieve the well-guided form of living organisms. To address them, it is important to consider the influence of the developmental process on the overall shape. Besides, the underlying constraints that this development imposes locally to the cells must be considered. The idea of the algorithms we have developed to study this influence is greatly inspired by the concept that Waddington introduced in 1940 as *the epigenetic landscape* [20]. The barriers through this landscape are similar to the constraints that the form faces through its development. Thus, each possible path of the landscape represents a possible form evolution process. In fact, an evolution is held in given conditions where only some specific genes can be activated (see Fig. 6). The identification of the obstacles across the landscape allows to know where and when feedback mechanisms are involved (i.e. differentiation that gives a particular character to the form at this time).

To study the effect of the developmental process constraints, we verify the conditions under which feedback mechanisms are triggered. To do so, we have

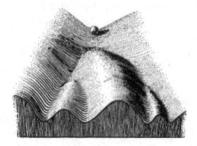

**Fig. 6.** In 1957, Conrad Waddington proposed the concept of an epigenetic landscape to represent the process of cellular decision-making during development. At various points in this dynamic visual metaphor, the cell (represented by a ball) can take specific permitted trajectories, leading to different outcomes or cell fates [8].

generated a base of genomes formed from the same set of genes. Having a genome that achieves a given shape, we aim to explore the entire space of possible genomes. The aim is to see if we can find another genome which allows to reach it.

In this paper we have chosen to target the well known *French flag*. That means we have sought genomes that give forms with these following characteristics:

– activation of the same genes as the *French flag* (same colors),
– choice of same segmentation directions as the *French flag* (same shape).

In Waddington's point of view, it implies that from different origins on the top of the *epigenetic landscape*, the ball can go through different trajectories and finally hits the same endpoint.

## 4.1  Software Architecture

In our software architecture, we have two main classes which handle exploration and simulation activities.

1. `Genome`: represents a genome and allows us to manage all treatments that can be performed on a genome. As we designed it, a genome is composed of one or more genes that are activated or not during the simulation.
2. `GenerateGenome`: includes all the activities of exploration, initialization, launching and stop. It carries initialization of the exploration's parameters methods, start and stop genome's simulation methods, genomes' treatments methods, set of genomes' managing methods, exploration results' recording methods.

## 4.2  Genomes Base Construct

Genes are defined in an `xml file` with all their attributes (*coded color, applied route choice, number of repetitions in the genome*, etc.). From this file, they are loaded into a `vector`. And from this vector, we create all possible combinations of genomes. Their construction is based on the characteristics of the *French flag* genome which are:

1. must be composed of 4 genes
2. the 1st gene in the genome must be the same as the 4th gene
3. if n is the number of times in which is repeated the 1st gene in the genome (called *multiplicity* of the gene), the 2nd gene must have therefore a *multiplicity* of 2xn and n for the 3rd and 4th genes.

Thus, all genomes are constructed in the same way (see Table 1).

**Table 1.** *French Flag* genome characteristics

| Genes | Gene 1 | Gene 2 | Gene 3 | Gene 1 |
|-------|--------|--------|--------|--------|
| Multiplicity | n | 2xn | n | n |

**Table 2.** *French Flag* genome simulation results

| Multiplicity | Number of cells | Age |
|--------------|-----------------|-----|
| 3 | 70 | 1.16 |
| 4 | 117 | 1.224 |
| 5 | 176 | 1.288 |
| 6 | 247 | 1.352 |

### 4.3   Initialization

During initialization, we define matching criteria to be achieved by simulated genomes in the base. When simulating *French flag* genome, we chose 3 as the first gene's order of *multiplicity*. We note that the multiplicity doesn't affect the shape but just its size. The bigger it is, the larger is the shape. And 3 is an optimal value for an exploration of such a large base.

In Table 2, we have some informations from different simulations performed with *French flag* genome. For a given *multiplicity* of the *French flag's* genome (row 1), we have the values of the criteria recorded during the simulation. The *number of cells* that make it up (row 2) and the *age* at which it reaches its final form (row 3). The current time is determined during the simulation by the scheduler and is expressed as a double.

So, during the exploration, to assess that a given genome has achieved the *French flag*, we mainly refer to these values of these criteria.

### 4.4   Description of the Algorithm

The goal of the algorithm is to explore all genomes. Each one is simulated in order to check if it is able to reach the *French flag*. In other terms, if it can achieve the same number of cells as the *French flag* in an identical time. If a such genome is found, the algorithm lists it as well as the order of *multiplicity* with which it has converged.

Outside a match, there may be two cases (see Fig. 7):

1. the number of cells exceeds that corresponding to the current choice of *multiplicity*. So we definitely assume that we can't reach the *French flag* with this genome. In this case, the algorithm continues exploring and pick its follower in the base.
2. the number of cells doesn't reach that of the current choice of *multiplicity*. So, the algorithm reloads and simulates the same genome, but leads it to a higher order of *multiplicity* (i.e. from 3 to 4). In the 2nd simulation of the

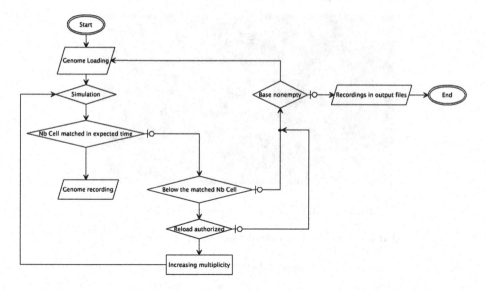

**Fig. 7.** The algorithm's flowchart

genome, if the number of cells corresponding to this new choice of *multiplicity* is met, then the genome is finally listed. Else, if the number of cells exceeds, we execute the same treatment as in (**1**). Else, if during this second simulation of the genome, the corresponding number of cells is still not reached, we go back to (**2**): increase *multiplicity* from **4** to **5** in the **3rd** simulation of the genome.

While the number of cells keeps being lower than that expected, we redefine the current genome by modifying its genes' parameters of *multiplicity*. This operation is repeated a maximum allowed number of times. In the algorithm, it is 3 times. So the last choice of *multiplicity* is 6 with an expected number of cell of 247 in 1.352 time units. In this step, the genome is no longer reloaded, considered definitely not able to achieve the *French flag*. Then, the algorithm continues exploring and picks its follower in the base.

When the algorithm has finished simulating all genomes, the program stops and we retrieve the results in an output file.

# 5    Evaluation

## 5.1    Test Conditions

In the tests we have achieved, the targeted pattern was, as shown in Fig. 8, the *French flag* introduced by L. Wolpert in 1969 [21].

The genome which allows to generate the *French flag* is as follows (details in Table 3): [1 **2 3** 1].

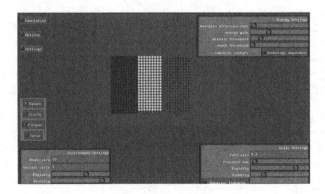

**Fig. 8.** Pattern targeted by genomes in the base

**Table 3.** Informations carried by the *French flag* genome

| Genes | Colour coded | Direction indicated in the route choice |
|-------|--------------|------------------------------------------|
| 1     | Blue         | Right                                    |
| 2     | White        | Forward                                  |
| 3     | Red          | Left                                     |

For the 64 genes we have, we obtained 262,114 possible combinations of genomes. This number made the exploration space pretty huge. Indeed, the simulation of the entire base took 5 days. Simulations were done on a computer with the following characteristics:

– System: Linux 32 bits
– Processor: Intel Core 2 CPU 6300 1.86GHz x 2
– RAM: 2 Go
– GPU: ATI Readon X1300/X1550 Series (256MB)

## 5.2   Output Results

5 genomes were identified as generating forms with characteristics identical to those of the *French flag* (see Fig. 9):

$$[1\ 2\ 4\ 1] - [1\ 4\ 2\ 1] - [3\ 2\ 1\ 3] - [4\ 1\ 2\ 4] - [4\ 3\ 2\ 4].$$

They have all achieved it with the first order of multiplicity (i.e. 3). Consequently, they are each composed of 70 cells with a maturation age of 1.16 time units (ref. Table 2).

We note that among these 5 forms retained, there are only 2 in which exactly the three genes of the *French flag* are activated: [4 3 2 4]-[3 2 1 3]. And only 1 of these 2 is same as *French flag*: [3 2 1 3].

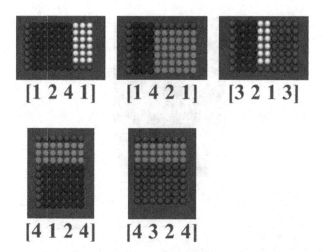

[1 2 4 1]     [1 4 2 1]     [3 2 1 3]

[4 1 2 4]     [4 3 2 4]

**Fig. 9.** Forms identified by the algorithm that have met matching criteria of the *French flag*

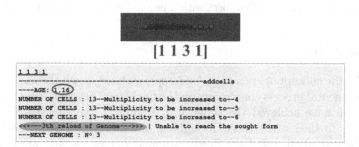

[1 1 3 1]

**Fig. 10.** Example of a discarded form after three attempts of reaching the *French flag*

So, from two different genomes, we have reached the same form.

Two generated forms discarded by the algorithm are presented in Figs. 10 and 11.

The first one presents the case where removal is due to the fact that the algorithm has reloaded the 3 times authorized the same genome [1 1 3 1] without reaching the sought pattern. Based on the output results of the simulation, we can point out that for these 3 times, the number of cells remains 13. Then, the algorithm moves to the next genome.

The second one presents the case where removal is due to the exceed number of cells in the generating form. An excess relative to that of the sought pattern with respect to the defined order of *multiplicity*. Here too, output results show that the algorithm aborts the simulation of the genome [3 2 12 3] and proceeds to the next one in the base.

[3 2 12 3]

```
3 2 12 3
-----------------------------------------------addcells
----AGE: (1.16)
--NUMBER OF CELLS : 80--Overtaking the right number 70
---NEXT GENOME : N° 8269
```

**Fig. 11.** Example of a discarded form with a number of cells exceeded that targeted

## 6    Conclusion

Inspired by the concept of *epigenetic landscape*, we have shown the importance of the developmental process constraints on the final form. Starting from 2 different genomes, we have achieved the same form thanks to the choices of feedback mechanisms to deal with stress throughout development. Theses mechanisms are, on the one hand, cellular differentiation. It allows the activated genes to encode the colors of the form. On the other hand, feedback mechanisms imply successive changes of direction. This allows to generate (to design) the shape as it goes along.

### 6.1    Relevance to Biological Issues

The model is tested in French flag pattern because it is a widely known issue in studying the factors that put into play in cells' differentiation. It constitutes the first steps towards the comprehension of biological mechanisms. Nevertheless, in future works, it might be also possible to test the model in real case. For example, in genetics, *single nucleotide polymorphism* (SNP) is the variation of a single base pair of the genome between individuals of same species. They are used for recognition of individuals or reconstruction of family trees. In our algorithm, we could build the genomes' base so that they are all identical except the last gene in each of them. This gene, different from one genome to another, would constitute our SNP. Then, to study the influence of this SNP, we can simulate the genomes and compare the generated forms.

We can also consider applying the algorithm to look for genomes that can generate more complex forms. Indeed, from the cell lineage of an organism, we can know its number of cells and its different type of cells. We can take the case of Caenorhabditis Elegans (*C. Elegans*). This is an eukaryote[2], which means that it shares cellular structures, molecular structures and control channels with organisms of higher species. Thus, its biological informations (embryogenesis, morphogenesis, growth, etc.) could be directly applicable to more complex organisms, such as humans. In addition, it has a fixed number of cells. The adult hermaphrodite is composed of 959 somatic nuclei and the adult male of 1031 while the young consists of 1090. The algorithm can be used to search in the base a genome which can generate a form with the following characteristics: an identical number of cells as the *C. Elegans* and the same number of activated genes than cellular types in *C. Elegans*.

## 6.2   Future Works

To consider the description of all types of dynamics, we plan to study a generalization of our mathematical model. In the current model, when a cell initiates mitosis, we just create next to it another one. The mother cell does not implement any other mechanism. Therefore, we will consider a mathematical model where the mother cell can in each time step:

- remain quiescent as it is in the current model,
- migrate to another point or,
- die.

This formalization will allow to reach all possible forms. The development of a such model raises new challenges in mathematical and numerical perspectives. Because describing all possible dynamics for each cell at each time step offers many options to consider. In computing, *parallel programming* on GPU[3] could overcome scalability and duration of simulations.

# References

1. Aubin, J.: Mutational and Morphological Analysis: Tools for Shape Regulation and Morphogenesis. Birkhauser, Boston (2000)
2. Aubin, J.-P.: Viability Theory. Birkhauser, Boston (1991)
3. Chevaillier, P., Bonneaud, S., Desmeulles, G., Redou, P.: Experimental study of agent population models with a specific attention to the discretization biases. In: Proceedings of the European Simulation and Modelling Conference (ESM'09), Leicester, UK, pp. 323–331 (2009)
4. Desmeulles, G., Querrec, G., Redou, P., Misery, L., Rodin, V., Tisseau, J.: The virtual reality applied to the biology understanding: the in virtuo experimentation. Expert Syst. Appl. **30**(1), 82–92 (2006)

---

[2] A eukaryote is an organism whose cells contain complex structures enclosed within membranes.

[3] Graphics Processing Unit.

5. Doursat, R.: Organically grown architectures: creating decentralized, autonomous systems by embryomorphic engineering. In: Würtz, R.P. (ed.) Organic Computing. Springer, Heidelberg (2008)

6. Fronville, A., Harrouet, F., Desilles, A., De Loor, P.: Simulation tool for morphological analysis. In: Proceedings of the European Simulation and Modelling Conference (ESM'2010), Hasselt, Belgium, pp. 127–132 (2010)

7. Fronville, A., Sarr, A., Ballet, P., Rodin, V.: Mutational analysis-inspired algorithms for cells self-organization towards a dynamic under viability constraints. In: SASO 2012, 6th IEEE International Conference on Self-Adaptive and Self-Organizing Systems, Lyon, France, pp. 181–186 (2012)

8. Goldberg, A.D., Allis, C.D., Bernstein, E.: Epigenetics: a landscape takes shape. Cell **128**, 635–638 (2007)

9. Henderson, J., Carter, D.: Mechanical induction in limb morphogenesis: the role of growth-generated strains and pressures. Bone **31**(6), 645–653 (2002)

10. Kauffman, S.: The Origins of Order: Self-Organization and Selection in Evolution. Oxford University Press, New York (1993)

11. Lawson, B., Park, S.: Asynchronous time evolution in an artificial society mode. J. Artif. Soc. Soc. Simul. **3**(1) (2000)

12. Lorenz, T.: Mutational Analysis - A Joint Framework for Cauchy Problems In and Beyond Vector Spaces. Springer, Berlin (2010)

13. Melani, C., Peyriéras, N., Mikula, K., Zanella, C., Campana, M., Rizzi, B., Veronesi, F., Sarti, A., Lombardot, B., Bourgine, P.: Cells tracking in the live zebrafish embryo. Conf. Proc. IEEE Eng. Med. Biol. Soc. **1**, 1631–1634 (2007)

14. Müller, G., Newman, S.: Origination of Organismal Form: Beyond the Gene in Developmental and Evolutionary Biology. MIT Press, Cambridge (2003)

15. Pena, A.C.: Un modèle de développement artificiel pour la génération de structures cellulaires. Ph.D. thesis, Université de Toulouse, décembre 2007

16. Peyriéras, N.: Morphogénèse : L'origine des formes, chapter Morphogénèse animale, pp. 179–201. Belin, Paris (2006)

17. Reignier, P., Harrouet, F., Morvan, S., Tisseau, J., Duval, T.: ARéVi: a virtual reality multi-agent platform. In: Heudin, J.-C. (ed.) Virtual Worlds 98. LNCS (LNAI), vol. 1434, pp. 229–240. Springer, Heidelberg (1998)

18. Southern, J., Pitt-Francisb, J., Whiteleyb, J., Stokeleyc, D., Kobashid, H., Nobesa, R., Kadookad, Y., Gavaghan, D.: Multi-scale computational modelling in biology and physiology. Prog. Biophys. Mol. Biol. **96**(9), 60–89 (2008)

19. Varela, F.: Principles of Biological Autonomy. North- Holland, New York (1979)

20. Waddington, C.: Organisers and Genes. CambridgeUniversity Press, Cambridge (1940)

21. Wolpert, L.: Positional information and the spatial pattern of cellular differentiation. J. Theor. Biol. **25**, 1–47 (1969)

# Biostatistics Regular Session

# High–Dimensional Sparse Matched Case–Control and Case–Crossover Data: A Review of Recent Works, Description of an R Tool and an Illustration of the Use in Epidemiological Studies

Marta Avalos[1,2,3](✉), Yves Grandvalet[4], Hélène Pouyes[1,5], Ludivine Orriols[1,2], and Emmanuel Lagarde[1,2]

[1] INSERM U897-Epidémiologie-Biostatistique, 33000 Bordeaux, France
[2] Univ. Bordeaux, 33000 Bordeaux, France
[3] INRIA SISTM Bordeaux, 33000 Bordeaux, France
marta.avalos@isped.u-bordeaux2.fr
[4] CNRS, Heudiasyc UMR6599, Univ. Compiègne, 60200 Compiègne, France
[5] Univ. de Pau et des Pays de l' Adour, 64012 Pau, France

**Abstract.** The conditional logistic regression model is the standard tool for the analysis of epidemiological studies in which one or more cases (the event of interest), are matched with one or more controls (not showing the event). These situations arise, for example, in matched case–control and case–crossover studies. In sparse and high-dimensional settings, penalized methods, such as the Lasso, have emerged as an alternative to conventional estimation and variable selection procedures. We describe the R package clogitLasso, which brings together algorithms to estimate parameters of conditional logistic models using sparsity-inducing penalties. Most individually matched designs are covered, and, beside Lasso, Elastic Net, adaptive Lasso and bootstrapped versions are available. Different criteria for choosing the regularization term are implemented, accounting for the dependency of data. Finally, stability is assessed by resampling methods. We previously review the recent works pertaining to clogitLasso. We also report the use in exploratory analysis of a large pharmacoepidemiological study.

**Keywords:** Software tool · Biostatistics · Algorithms for pharmacoepidemiology · Penalized conditional logistic regression · Lasso

## 1 Introduction

Epidemiological case–control studies are used to identify factors that may contribute to a medical condition by comparing a group of cases, that is, people with the condition under investigation (a disease, for example), with a group of controls who do not have the condition but who are believed to be similar

© Springer International Publishing Switzerland 2014
E. Formenti et al. (Eds.): CIBB 2013, LNBI 8452, pp. 109–124, 2014.
DOI: 10.1007/978-3-319-09042-9_8

in other respects. Logistic regression is the most important statistical method in epidemiology to analyze data arising from a case-control study. It allows to account for the potential confounders and, if the logistic model is correct, to eliminate their effect.

Cases and controls are sometimes matched: every case is matched with a preset number of controls who share a similar exposure to these matching factors, to ensure that controls and cases are similar in variables that are related to the variable under study but are not of interest by themselves [1]. Matching is useful when the distributions of the confounders differs radically between the unmatched comparison groups. In these situations, the weight of confounding factors is so important that a simple adjustment does not guarantee a straightforward interpretation of results. Advantages (as elimination of confounding or gain in efficiency) and disadvantages (as possible reduction in sample size or introduction of new bias if the matching factor is not a confounder but is in the causal pathway between exposure and disease), use and misuse of matching have been widely discussed in the literature [1–10].

The case-crossover design, in which each subject serves as his own control, is a particular matched case-control design [11,12]. The association between event onset and risk factors is estimated by comparing exposure during the period of time just prior to the event onset (case period) to the same subject's exposure during one or more control periods. As a result, this design inherently eliminates the bias in control selection and removes the confounding effects of time-invariant factors. However, the case-crossover design is sensitive to the effects of time-varying risk factors [13,14].

The conditional logistic regression model is the standard tool for the analysis of individually matched case–control and case–crossover studies. Usually, regression coefficients are estimated by maximizing the conditional log–likelihood function and variable selection is performed by conventional manual or automatic selection procedures, such as stepwise. These techniques are, however, unsatisfactory to analyze high-dimensional data which arise nowadays in many diverse fields of epidemiological research (genetic, environmental or pharmaco–epidemiology, for instance).

Fortunately, appropriate statistical methods have been proposed to address problems in high-dimensional settings where the number of relevant predictors is expected to be small. The issue there is to find reasonably accurate sparse solutions that are easy to interpret and can be used for the prediction and/or estimation of the predictors effects on the response. Penalized methods, such as the Lasso (*least absolute shrinkage and selection operator*) [15], have emerged as an alternative. In particular, the Lasso and related methods have recently been adapted to conditional logistic regression [16,17].

## 2    Conditional Logistic Regression

In matched case-control, the data are not independent. Each case is seen as one stratum with its set of matched controls. The standard analysis tool is then conditional logistic regression [18], which differs from ordinary logistic regression in that it allows intercept to vary among the matched sets. As a consequence of matching, the effects of matching covariates on the response are not estimable. Also, the likelihood function has to be conditional on the design.

## 2.1   The Model

We focus our attention on the relationship between several input variables $X = (X_1, \ldots, X_p)$ and the binary response variable $Y \in \{0, 1\}$. In the matched case-control design, observations are grouped into strata (indexed by $n = 1, \ldots, N$), and each stratum consists in one case (with $y = 1$) and $M$ controls (with $y = 0$). The vector of observations for subject $i$ of stratum $n$ is $\mathbf{x}_{in} = (x_{in1}, \ldots, x_{inp})$, $i = 1, \ldots, M+1$. The data matrix of dimension $(N \times (M+1)) \times p$ is denote $\mathbf{X}$.

Let $P_{in}$ be the (unconditional on the design) probability of occurrence of the event for the $i$-th subject of the $n$-th stratum. Consider the logistic model, supposing that disease risk differs among strata:

$$P_{in} = P(Y_{in} = 1|\mathbf{x}_{in}) = \frac{1}{1 + e^{-(\alpha_n + \mathbf{x}_{in}\boldsymbol{\beta})}}, \tag{1}$$

where $\alpha_n$ are coefficients representing the effect of matching covariates on the response; and coefficients $\boldsymbol{\beta} = (\beta_1, \ldots, \beta_p)'$ represent the effect of predictors.

## 2.2   Conditional Likelihood

When the differences among strata are not of interest, we only wish to estimate $\boldsymbol{\beta}$. Consider the $n$-th stratum, the unconditional probability of observing the occurrence of the event only in the $i$-th subject is:

$$(1 - P_{1n}) \ldots (1 - P_{i-1n}) P_{in}(1 - P_{i+1n}) \ldots (1 - P_{M+1n}). \tag{2}$$

Under the logistic model, the conditional probability (conditional on the design where each stratum consists in 1 case and $M$ controls) is given by:

$$\frac{\frac{P_{in}}{1-P_{in}} \prod_{l=1}^{M+1}(1 - P_{ln})}{\sum_{l=1}^{M+1} \frac{P_{ln}}{1-P_{ln}} \prod_{i=1}^{M+1}(1 - P_{in})} = \frac{e^{\mathbf{x}_{in}\boldsymbol{\beta}}}{\sum_{l=1}^{M+1} e^{\mathbf{x}_{ln}\boldsymbol{\beta}}}. \tag{3}$$

Using the convention that all cases are indexed by $l = 1$ ($Y_{1n} = 1$ and $Y_{ln} = 0$, $l \neq 1$), the negative log conditional likelihood loss function, evaluated at $\boldsymbol{\beta}$ and $D = \{(\mathbf{x}_{in}, y_{in})\}_{i=1,\ldots,M+1;n=1,\ldots,N}$ can be written as:

$$L(\boldsymbol{\beta}, D) = -\sum_{n=1}^{N} \left[ \mathbf{x}_{1n}\boldsymbol{\beta} - \log \left( \sum_{l=1}^{M+1} e^{\mathbf{x}_{ln}\boldsymbol{\beta}} \right) \right]. \tag{4}$$

Usually, the estimation of the parameters vector $\boldsymbol{\beta}$ is found by minimizing the negative log conditional likelihood loss function:

$$\widehat{\boldsymbol{\beta}}_D^{\text{ML}} = \underset{\boldsymbol{\beta}}{\operatorname{argmin}} \left( L(\boldsymbol{\beta}, D) \right), \tag{5}$$

and inference is based on the estimated covariance matrix of the parameter estimates. Notice that $\alpha_n$ disappear from the likelihood function.

## 2.3  $L^1$ Penalized Conditional Likelihood

The penalization of likelihoods by $L^1$-norms has become an established and relatively standard technique with sparse high-dimensional data: these techniques may improve prediction accuracy (since regularization leads to variance reduction) together with interpretability (since sparsity identifies a subset of variables with strong effects). Computationally, these penalties are attractive (the optimization problem is convex), and their theoretical properties have been intensively studied during the last years.

The Lasso applied to conditional logistic regression consists in minimizing the negative log conditional likelihood penalized by the $L^1$-norm of the unknown coefficient vector:

$$\widehat{\boldsymbol{\beta}}_D^{\mathrm{L}}(\lambda) = \underset{\boldsymbol{\beta}}{\operatorname{argmin}} \left( L(\boldsymbol{\beta}, D) + \lambda \|\boldsymbol{\beta}\|_1 \right),$$

where $\lambda$ is a regularization parameter, and $\|\boldsymbol{\beta}\|_1 = \sum_{j=1}^{p} |\beta_j|$ is the $L^1$-norm of coefficients. Consistency properties of Lasso for estimating the regression parameter and for variable selection have been well studied in the linear regression case (see, for example, [19–25]) and, in a second time, in the logistic regression case (see, for example, [26–29]). While the Lasso has excellent properties in dimensionality reduction and estimation, it generally does not lead to consistent model selection, only to conservative model selection: with a proper penalty parameter, the Lasso retains the relevant variables, but also a few additional irrelevant ones (see for example, [30] and [31] for logistic regression).

Different procedures have been proposed to address this particular problem. In the adaptive Lasso [32], weights are incorporated in the $L^1$-norm to provide an adaptive penalization of coefficients:

$$\widehat{\boldsymbol{\beta}}_D^{\mathrm{AL}}(\lambda) = \underset{\boldsymbol{\beta}}{\operatorname{argmin}} \left( L(\boldsymbol{\beta}, D) + \lambda \sum_{j=1}^{p} w_j |\beta_j| \right).$$

The adaptive Lasso attempts to reduce the estimation bias and improve variable selection accuracy by assigning heavy penalties to zero coefficients and lighter ones for nonzero coefficients. The weights are often chosen to be inversely proportional to the magnitude of a basic estimate of the coefficients, such as the maximum likelihood estimate $w_j = |\widehat{\beta}_j^{\mathrm{ML}}|^{-1}$ or the Lasso estimate $w_j = |\widehat{\beta}_j^{\mathrm{L}}|^{-1}$, with the convention $w_j = \infty$ if $|\widehat{\beta}_j^{\mathrm{L}}| = 0$ (if a predictor is not selected by the Lasso, it will also not be selected by the adaptive Lasso [27]). The latter choice is operative when the total number of predictors is larger than the sample size.

Other enhancements of the Lasso based on resampling techniques have been introduced in the literature, starting with Bolasso [22], which consists in drawing bootstrap samples from the original training set and then intersecting the supports of the Lasso bootstrap estimates or, in the soft version of Bolasso, selecting predictors frequently present in the bootstrap replications (see also [33–36]).

The random Lasso [37] randomly selects candidate predictors and then averages the predictive models adjusted on bootstrap samples. This procedure alleviate another known weakness of the Lasso, which tends to select only few of the relevant predictors when the latter are highly correlated. The elastic net [38] is another proposal targeting the same issue, using a deterministic approach and a different penalty:

$$\widehat{\boldsymbol{\beta}}_D^{\text{EN}}(\lambda_1, \lambda_2) = \operatorname*{argmin}_{\boldsymbol{\beta}} \left( L(\boldsymbol{\beta}, D) + \lambda_1 \|\boldsymbol{\beta}\|_1 + \lambda_2 \|\boldsymbol{\beta}\|_2 \right), \tag{6}$$

where the Lasso penalty, controlled by the regularization parameter $\lambda_1$, is complemented with a quadratic penalty term controlled by $\lambda_2$, designed to compensate for the correlation between predictors.

## 3    Complexity Tuning

The parameter $\lambda \geq 0$ controls the complexity of the model, so that if $\lambda \to \infty$, we obtain the "null model" in which no variable is useful, whereas if $\lambda = 0$, the solution is the unpenalized conditional logistic regression estimate obtained by minimizing the likelihood-based loss function. The performance of the Lasso and related methods depends crucially on the choice of the tuning parameter [28, 39].

When prediction accuracy is sought, $\lambda$ can be estimated by cross-validation. The data set $D$ is first chunked into $K$ disjoint blocks of the same size $N/K$, where $N$ is the number of strata (we recall here that strata correspond to subjects and assume to simplify that $N$ is a multiple of $K$). Let us write $D_k$ for the $k$-th block, and $D \backslash D_k$ the training set obtained by removing the elements in $D_k$ from $D$. The negative conditional log-likelihood-based cross-validation criterion, preserving the matching of data, is [40, 41]:

$$CV(\lambda) = \frac{1}{K} \sum_{k=1}^{K} L\left(\widehat{\boldsymbol{\beta}}_{D \backslash D_k}(\lambda), D_k\right).$$

$\widehat{\lambda} = \operatorname{argmin}_\lambda CV(\lambda)$ minimizes the *average* of the log-likelihood, evaluated on test block $D_k$, of the parameters estimated from the training set deprived from $D_k$.

Common choices for $K$ are 5 and 10. Note that $K$-fold cross-validation with $K = N$ is leave-one-out cross-validation, recommended for small sample sizes if the signal-to-noise ratio is large enough [42]. Leave-one-out cross-validation can be approximated by generalized cross-validation:

$$GCV(\lambda) = \frac{2L(\boldsymbol{\beta}, D)}{[1 - \text{df}(\lambda)/(N \times (M+1))]^2},$$

where $\text{df}(\lambda)$ corresponds to the effective number of parameters [43]. A simple unbiased estimator of $\text{df}(\lambda)$ is the number of coefficients estimated to be nonzero [44, 45]. For large datasets, we can use the Akaike information criterion (AIC):

$$AIC(\lambda) = 2L(\boldsymbol{\beta}, D) + \text{df}(\lambda)2\big(N \times (M+1)\big).$$

AIC is efficient (asymptotically optimal) and it is asymptotically equivalent to leave-one-out cross-validation, without incurring the additional computational cost.

Cross-validation does not primarily target the identification of the set of truly relevant variables (also known in this context as support recovery), since prediction and identification are somehow conflicting goals [46–48]. A voting-based cross-validation in which $\widehat{\lambda}$ is the $\lambda$-value that minimizes a *majority* of the likelihood-based loss function evaluated at a given test block when the training set is deprived from him is model consistent when a larger proportion of data is put in the test block [47, 49]. Thus, $\widehat{\lambda}$ is the more frequent value in $\{\widehat{\lambda}_k\}_{k=1,\dots,K}$, where:

$$\widehat{\lambda}_k = \text{argmin}_\lambda L\big(\widehat{\beta}_{D_k}(\lambda), D\backslash D_k\big).$$

When the Lasso is primarily used to identify relevant variables, the Bayesian information criterion (BIC) may be appropriate:

$$BIC(\lambda) = 2L(\beta, D) + \text{df}(\lambda) \log \big(N \times (M + 1)\big).$$

Other criteria aim to control the false discovery rate [50].

Two parameters have to be tuned in the elastic net (6) and the soft version of Bolasso ($\lambda$ and the frequency threshold). A two–dimensional grid search is recommended for the elastic net (instead of successive one-dimensional optimization) [51]. Knowledge-driven [52, 53] as well as data-driven (CV, AIC) [54, 55] approaches are used to choose the frequency threshold of Bolasso.

## 4    Uncertainty Measures

Lasso-related methods are useful for constructing analytic models in high–dimensional settings. However, recent reviews have pointed out the under-use of these new techniques in epidemiological publications [56, 57], despite admitted drawbacks of stepwise-type procedures and other conventional methods [58–61].

We believe that the main obstacle for the use of penalization techniques in epidemiological studies resides in the lack of undisputed uncertainty measures attached to estimators. Confidence intervals based on approximations of the covariance matrix of the estimated coefficients have been derived [15, 32, 62, 63], but their reliability is questionable [19, 64]. Few of the studies that use Lasso-related methods to analyze real data provide measures of uncertainty. These few studies use bootstrapping to generate measures of statistical accuracy. Reference [65] use the mean and the standard error of the estimates computed from the bootstrap samples, and the percentage of the bootstrap coefficients at zero to validate the chosen coefficient estimates. Reference [66] use bootstrap-normal confidence intervals in the analysis of genetic data to indicate the statistical importance of a selected variable. Reference [53] complement the Lasso step by a bootstrap step to measure the frequency with which each predictor is chosen by the Lasso procedure.

This is a critical problem as it is known that confidence intervals of sparse estimators are larger than the ones of ordinary likelihood estimators, when the tuning parameter is chosen to lead to consistent model selection [67–69]. The discontinuity of the sampling distribution of the parameters raises also a difficulty regarding the interpretation of standard errors [70]. Some authors consider that standard errors are not very meaningful for penalized estimates since the variance of estimators is reduced by introducing substantial bias: the bias is then a major component of the mean squared error of the estimator, while the variance may only contribute minimally. As a result, bootstrap would give an assessment of the variance of the estimates, but could not give a sufficiently accurate estimate of the bias [71]. Bootstrap-based confidence intervals have been shown to be inconsistent in the presence of any irrelevant variable (true zero coefficients) [19, 64]. The residual bootstrap mimics the main features of the regression model closely but it fails to reproduce the sign of the zero-components of $\beta$ with sufficient accuracy in the formulation of the bootstrap Lasso estimation criterion, leading to a random limit, and subsequently, to inconsistency.

Reference [36] proposed a modified version of the residual based bootstrap method for linear regression that takes into account the particularities of the limiting distribution of the $L^1$-penalized least squares estimator. The idea is to force components of the Lasso estimator $\widehat{\beta}$ to be exactly zero whenever they are close to zero. The modified bootstrap is consistent for estimating the distribution of the estimator, however it is not uniformly consistent and so the resulting confidence interval will most likely not be valid uniformly with respect to the unknown parameter values. Reference [72] adapted this method to the $L^1$-penalized logistic model by different bootstrap procedures (standard residual bootstrap, one-step residual resampling based on the ridge-based reformulation of Lasso, double bootstrap). Reference [73] used bootstrap percentile confidence intervals to account for uncertainty in selection by adapting the modifications suggested in [36] to logistic regression.

At the moment it is still not clear how to construct valid confidence intervals for this type of estimators. Reference [68] have shown how to do it in the case of linear orthogonal design (asymptotically and also for finite samples) but adjusting it for more general situations is still an open research topic. Another possibility is to adopt the Bayesian formulation of Lasso that naturally leads to valid standard errors [74, 75].

## 5    Standardization, Bias Correction, Unpenalized Predictors

Penalization techniques based on norms being sensitive to scaling, covariates are usually standardized (and, sometimes, also mean-centered), in order to compare and remove predictors from the model. It should be noted that, in an identification approach, even if predictors are in the same units already (as in the case of binary covariates), it is useful to standardize: without scaling, the Lasso estimate has a tendency to disregard predictors with small variability on the sample used for estimation which may be as important regarding inference as more commonly ones. A classical scaling approach in penalized regression model is from 0 to 1. This would prevent detrimental effects on binary covariates which would be actually unscaled. Covariates can be standardized by dividing by their respective standard errors or the diagonal elements of the covariance matrix of the minimum likelihood estimator [76].

Penalized estimation goes through the introduction of a bias on the estimated coefficients. To correct bias, we can fit the unpenalized regression model with the covariates retained in the model (those having a nonzero point estimate) [77].

A common approach in epidemiological studies consists in forcing some of the covariates into the model in order to ensure that the apparent differences between predictor values of cases and controls are not misleadingly created by confounding covariates. Thus, to control for potential confounding, these covariates can be unpenalized, so that they are not permitted to drop out of the model.

Usually, in epidemiological studies, categorical variables are coded using grouped dummy variables and pairwise-products of original variables representing pairwise-interactions are included in the model. In that cases, to enhance interpretability, grouped dummy variables corresponding to a given categorical variable are removed altogether or selected all them, and optional variables associated to relevant interactions become mandatory. Several proposals specifically address the question of interactions [78] or categorical variables [79].

## 6    Capabilities of `clogitLasso`

The R package `clogitLasso` (available on request from authors) implements, for small to moderate sample sizes (less than 3000 observations), the algorithms discussed in [17], based on the stratified discrete-time Cox proportional hazards model and depending on the `penalized` package [71]. For large datasets, `clogitLasso` computes the efficient procedures proposed in [80], based on an IRLS (iteratively re-weighted least squares) algorithm and depending on the `lassoshooting` package [81]. Methods and model selection criteria reviewed in Sects. 2.3 and 3 are available.

In the situation of unbalanced strata in a 1:M study and in the situation in which there is M controls matched to more than one case in each stratum, only the algorithm based on the stratified discrete-time Cox proportional hazards model applies. The `penalized` package is optimized for scenarios with many covariates but does not handle a large amount of observations, then these situations can be analyzed only for small to moderate sized samples. Elastic net is available when calling `penalized` but unavailable in the `lassoshooting` package.

A basic nonparametric bootstrap allows us to provide the frequency with which each predictor is chosen by the Lasso. A heuristic algorithm combining two ideas: the modification by means of a threshold constant, introduced in [36], and the pre-selection of candidate variables by means of a measure of importance, used in [37], is also proposed (if the threshold constant is fixed to 0 and the number of candidate predictors if fixed to be greater or equal to the number of nonzero Lasso coefficients, the procedure consists in bootstrapping only the selected predictors). This procedure is also closed to iterated Lasso [27] except that Bolasso replaces the adaptive Lasso procedure in which weights are determined in a previous Lasso step. Residual or one-step (using IRLS last step) versions of bootstrap [72] are however not available.

When variable standardization is performed prior to fitting, the coefficients are returned on the original scale.

clogitLasso depends on the survival package to estimate regression coefficients using (unpenalized) conditional maximum likelihood, when a bias correction step is applied. Classic maximum likelihood estimation is also needed when using adaptive Lasso or univariate logistic regression (predictor by predictor) ad hoc method. We have no solution by now concerning the handling of grouped categorical variables. In the situation in which there is an interest on interaction effects, we suggest to apply a two step procedure: in step 1, clogitLasso is applied to original predictors; in step 2, clogitLasso is applied to the predictors selected in step 1, forcing them into the model, plus their interactions.

## 6.1   Example 1

To illustrate the use of clogitLasso in a 1:M matched case-control study with unbalanced strata and $N < p$, we consider the infertility after spontaneous and induced abortion data from the survival R package. Each of the 83 cases (infertile women) is matched to one or two controls (fertile women) on age, education and parity (the number of children to which a woman has given birth). There are two predictors, the number of prior induced abortions and the number of prior spontaneous abortions (0, 1, 2 or more). We add 1000 irrelevant variables independently Bernoulli distributed with a probability of 0.8.

Figure 1 shows coefficient estimates for each variable as a function of $\lambda$. The vertical line indicates the $\lambda$ value minimizing the 10-fold likelihood cross-validation criterion ($\widehat{\lambda} = 6.4$). The cross-validation function of $\lambda$ is represented at right. The Lasso with a cross-validation criterion selected the two risk factors and some noise variable (9 among 1000, in the present example). As noticed above, when a prediction–based method, such as cross-validation, is used to estimate the regularization parameter, in general, the Lasso retains the relevant predictors, but also a few additional irrelevant ones, though, typically, their estimates are small. Indeed, the two truly associated variables have a positive coefficient and some of the 9 noise variables have negative signs, then the overall prediction of the response is slightly better than predicting the response by using only the two truly associated variables with lower coefficient values (for example, with $\widehat{\lambda} = 8.2$).

## 6.2   Example 2

To illustrate the use of clogitLasso in a large 1:1 case-crossover study, we consider the exposure to medicinal drugs and the risk of being involved in an

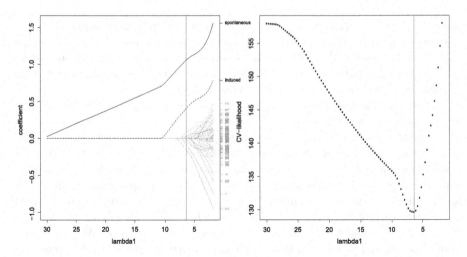

**Fig. 1.** $L^1$-regularization path as a function of $\lambda$ for the infertility data with 100 noise variables (left). Original predictors (the number of prior induced abortions and the number of prior spontaneous abortions) are plotted in black while noise variables are in grey. Cross–validation values as a function of $\lambda$ (right). The red vertical lines indicate the $\lambda$ value minimizing the 10-fold likelihood-based cross-validation criterion (Color figure online).

injurious non-alcohol related road traffic crash in France [55]. We consider all dispensed and reimbursed medicines to the drivers, coded by the WHO ATC (World Health Organization Anatomical Therapeutic Chemical) classification fourth level system. This leads to 402 candidate binary predictors (exposure, coded 1, and unexposure, coded 0, to each medicinal drug). The case period is designated as the one-day period preceding the crash, and the control period is designated as the one-day period one month preceding the crash, to avoid any residual effect of an exposure in the control period on the case one. 36,612 person-periods contributed to the estimation.

The use of Lasso–related techniques is justified in this context as follows. First, regression models, with straightforward interpretation, are the most important statistical techniques used in analytical epidemiology. Thus, these techniques appear to be a good compromise between traditional and data-driven approaches since modeling is based on standard regression models, rather than a black–box. Second, controlling for potential confounding is an critical point in epidemiology, thus multivariate modeling approaches are preferable to separate univariate tests. Third, it is expected that that only few drugs will be truly associated with the risk of being involved in a road traffic crash, thus sparsity-inducing penalties seem to be appropriate. It is also expected that some of these relevant drugs will have a weakly strength of association, however, only predictors with effect sizes above the noise level can be detected using Lasso–related techniques. Nevertheless, this limitation is shared by any model selection method [21,53,82]. Although some drugs are usually prescribed together and correlation problems are possible, we observed only mild correlation, probably because the large sample size.

Our main goal is to identify prescription drugs associated with an increased risk of crash. Here we considered Bolasso, thus only exposures frequently chosen by Lasso over the bootstrap samples are selected, which improves results' stability. Two parameters have to be tuned: lambda and the frequency threshold. We used 10-fold likelihood-based cross-validation to estimate the penalization parameter (for each bootstrap sample) and AIC to estimate the frequency threshold (over 1000 bootstrap samples). Penalization is intended to control variance, at the cost of introducing bias in the estimated coefficients. Subsequent to the Bolasso procedure, we corrected for bias.

Figure 2 shows coefficient estimates for each variable as a function of $\lambda$ for one of the bootstrap samples (left). The mean±standard deviation number of selected drugs across the 1000 resamples was $157 \pm 14$. For the particular bootstrap sample in the figure, cross-validation leaded to a model containing 151 drugs. However only 32 among them were selected in at least 78 % (the frequency threshold estimated using AIC) of the bootstrap samples. Coefficient values of drugs stably selected are indicated in black, the others are in grey.

Some pharmacoepidemiological results were consistent with literature (for example, benzodiazepine hypnotics: N05CD, anxiolytics: N05BA, N05BC).

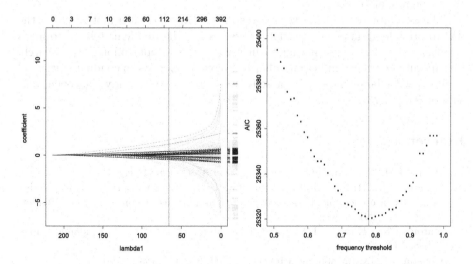

**Fig. 2.** $L^1$-regularization path as a function of $\lambda$ for one particular bootstrap sample of the case-crossover data of medicinal drugs related traffic accidents (left). The top axis indicates the number of drugs selected (with non null estimated coefficients) as a function of lambda. The red vertical line indicates the $\lambda$ value minimizing the 10-fold likelihood-based cross-validation criterion. Predictors frequently selected by Bolasso (in at least 78 % of the bootstrap samples) are plotted in black while the others are in grey. AIC values as a function of the frequency threshold (right). The red vertical line indicates the frequency value minimizing the AIC criterion (Color figure online).

Some results completed results of other studies (for instance, antiepileptics: N03AF, N03AG, N03AX, antidepressants: N06AB, N06AX, drugs used in opioid dependence: N07BC, and drugs used in diabetes: A10AD). Cardiovascular drugs: C10BA, C10BX presented relevant association with the risk of crash. These drugs are markers of cardiovascular diseases which themselves may increase the risk of accident involvement. Finally, antithrombotic agents: B01AB showed an inversed association, likely indicating that cardiovascular events may occur when the treatment is interrupted. This result oriented new researches.

# 7   Conclusion

The `clogitLasso` R package brings together penalized conditional likelihood methods for the analysis of sparse high-dimensional matched case-control data. The main underlying algorithms are presented in [17,80], they require both the `penalized` and the `lassoshooting` packages [71,81]. We have reviewed here the recent works pertaining to the `clogitLasso` package. Some theoretical properties of Lasso-type estimators are now well established, others are, however, still heuristic-based and the results obtained when applying that methods have to be interpreted with caution. This is notably the case for the validity of uncertainty measures and the consistency of model selection criteria when identification of the "true model" is sought.

The development of the `clogitLasso` R package has been motivated by the will to provide an appropriate tool for addressing the problems related to high-dimensionality that arise nowadays in many fields of epidemiological research. Our recent experience in helping pharmacoepidemiologists to conduct statistical analyses with large data from a self-matched case-only study has been very supportive [55,73].

# References

1. Bland, J.M., Altman, D.G.: Matching. BMJ **309**, 1128 (1994)
2. Kupper, L.L., Karon, J.M., Kleinbaum, D.G., Morgenstern, H., Lewis, D.K.: Matching in epidemiologic studies: validity and efficiency considerations. Biometrics **37**, 271–291 (1981)
3. Karon, J.M., Kupper, L.L.: In defense of matching. Am. J. Epidemiol. **116**, 852–866 (1982)
4. Constanza, M.C.: Matching. Preventive Med. **24**, 425–433 (1995)
5. Rothman, K., Greenland, S.: Modern Epidemiology, 2nd edn. Lippincott, Williams and Wilkins, Philadelphia (1998)
6. Stürmer, T., Brenner, H.: Flexible matching strategies to increase power and efficiency to detect and estimate gene-environment interactions in case-control studies. Am. J. Epidemiol. **155**, 593–602 (2002)
7. Vandenbroucke, J.P., von Elm, E., Altman, D.G., Gotzsche, P.C., Mulrow, C.D., Pocock, S.J., Poole, C., Schlesselman, J.J., Egger, M.: Strengthening the reporting of observational studies in epidemiology (strobe): explanation and elaboration. PLoS Med. **4**, 1628–1654 (2007)

8. Hansson, L., Khamis, H.: Matched samples logistic regression in case-control studies with missing values: when to break the matches. Stat. Methods Med. Res. **17**, 595–607 (2008)

9. Rose, S., Van der Laan, M.J.: Why match? investigating matched case-control study designs with causal effect estimation. Int. J. Biostat. **5**, Art. 1 (2009). doi:10. 2202/1557-4679.1127

10. Stuart, E.: Matching methods for causal inference: a review and a look forward. Stat. Sci. **25**, 1–21 (2010)

11. Maclure, M.: The case-crossover design: a method for studying transient effects on the risk of acute event. Am. J. Epidemiol. **133**, 144–153 (1991)

12. Delaney, J., Suissa, S.: The case-crossover study design in pharmacoepidemiology. Stat. Methods Med. Res. **18**, 53–65 (2009)

13. Mittleman, M., Maclure, M., Robins, J.: Control sampling strategies for case-crossover studies: an assessment of relative efficiency. Am. J. Epidemiol. **142**, 91–98 (1995)

14. Janes, H., Sheppard, L., Lumley, T.: Overlap bias in the case-crossover design, with application to air pollution exposures. Stat. Med. **24**, 285–300 (2005)

15. Tibshirani, R.: Regression shrinkage and selection via the lasso. J. Roy. Stat. Soc. Ser. B **58**, 267–288 (1996)

16. Avalos, M.: Model selection via the lasso in conditional logistic regression. In: Proceedings of the Second International Biometric Society Channel Network Conference, Ghent, Belgium, 6–8 April 2009

17. Avalos, M., Grandvalet, Y., Duran-Adroher, N., Orriols, L., Lagarde, E.: Analysis of multiple exposures in the case-crossover design via sparse conditional likelihood. Stat. Med. **31**, 2290–2302 (2012)

18. Breslow, N.E., Day, N.E.: Statistical Methods in Cancer Research. The analysis of case-control studies, vol. 1. IARC Scientific Publications, Lyon (1980)

19. Knight, K., Fu, W.: Asymptotics for lasso-type estimators. Ann. Stat. **28**, 1356–1378 (2000)

20. Zhao, P., Yu, B.: On model selection consistency of lasso. J. Mach. Learn. Res. **7**, 2541–2563 (2006)

21. Candes, E.J., Plan, Y.: Near-ideal model selection by L1 minimization. Technical report, Caltech, USA (2007)

22. Bach, F.: Bolasso: model consistent lasso estimation through the bootstrap. In: McCallum, A., Roweis, S.T. (eds.) Proceedings of the 25th International Conference on Machine Learning (ICML 2008), Helsinki, Finland, 5–9 July 2008

23. Zhang, T.: Some sharp performance bounds for least squares regression with L1 regularization. Ann. Stat. **37**, 2109–2114 (2009)

24. Wainwright, M.J.: Sharp thresholds for noisy and high-dimensional recovery of sparsity using L1-constrained quadratic programming (lasso). IEEE Trans. Inf. Theory **55**, 2183 (2009)

25. Juditsky, A., Nemirovski, A.: On verifiable sufficient conditions for sparse signal recovery via L1 minimization. Math. Program. **127**, 57–88 (2011)

26. Van de Geer, S.: High-dimensional generalized linear models and the lasso. Ann. Stat. **36**, 614–645 (2008)

27. Huang, J., Ma, S., Zhang, C.: The iterated lasso for high-dimensional logistic regression. Technical report, The University of Iowa, USA, No. 392 (2008)

28. Bunea, F., Barbu, A.: Dimension reduction and variable selection in case control studies via regularized likelihood optimization. Electron. J. Stat. **3**, 1257–1287 (2009)

29. Huang, J., Zhang, C.: Estimation and selection via absolute penalized convex minimization and its multistage adaptive applications. J. Mach. Learn. Res. **13**, 1839–1864 (2012)
30. Meinshausen, N., Yu, B.: Lasso-type recovery of sparse representations for high-dimensional data. Ann. Stat. **37**, 246–270 (2009)
31. Bach, F.: Self-concordant analysis for logistic regression. Electron. J. Stat. **4**, 384–414 (2010)
32. Zou, H.: The adaptive lasso and its oracle properties. J. Am. Stat. Assoc. **101**, 1418–1429 (2006)
33. Hall, P., Lee, E., Park, B.: Bootstrap-based penalty choice for the lasso, achieving oracle performance. Stat. Sinica **19**, 449–471 (2009)
34. She, Y.: Thresholding-based iterative selection procedures for model selection and shrinkage. Electron. J. Stat. **3**, 384–415 (2009)
35. Meinshausen, N., Bühlmann, P.: Stability selection. J. Roy. Stat. Soc. Ser. B **72**, 417–473 (2010)
36. Chatterjee, A., Lahiri, S.N.: Bootstrapping lasso estimators. J. Am. Stat. Assoc. **106**, 608–625 (2011)
37. Wang, S., Nan, B., Rosset, S., Zhu, J.: Random lasso. Ann. Appl. Stat. **5**(1), 468–485 (2011)
38. Zou, H., Hastie, T.: Regularization and variable selection via the elastic net. J. Roy. Stat. Soc. Ser. B **67**, 301–320 (2005)
39. Shi, W., Lee, K., Wahba, G.: Detecting disease-causing genes by lasso-patternsearch algorithm. BMC Proc. **1**(Suppl 1), S60 (2007)
40. Van der Laan, M., Dudoit, S., Keles, S.: Asymptotic optimality of likelihood-based cross-validation. Stat. Appl.Genet. Mol. Biol. **3**, Art. 4. (2004). doi:10.2202/1544-6115.1036
41. Van Houwelingen, H.C., Bruinsma, T., Hart, A.A.M., van't Veer, L.J., Wessels, L.F.A.: Cross-validated Cox regression on microarray gene expression data. Stat. Med. **25**, 3201–3216 (2006)
42. Arlot, S., Celisse, C.: A survey of cross-validation procedures for model selection. Stat. Surv. **4**, 40–79 (2010)
43. Ye, J.: On measuring and correcting the effects of data mining and model selection. J. Am. Stat. Assoc. **93**, 120–131 (1998)
44. Zou, H., Hastie, T., Tibshirani, R.: On the degrees of freedom of the lasso. Ann. Stat. **35**, 2173–2192 (2007)
45. Tibshirani, R.J., Taylor, J.: Degrees of freedom in lasso problems. Ann. Stat. **40**, 1198–1232 (2012)
46. Yang, Y.: Can the strengths of AIC and BIC be shared? a conflict between model identification and regression estimation. Biometrika **92**, 937–950 (2005)
47. Yang, Y.: Comparing learning methods for classification. Stat. Sinica **16**, 635–657 (2006)
48. Leng, C., Lin, Y., Wahba, G.: A note on the lasso and related procedures in model selection. Stat. Sinica **16**, 1273–1284 (2006)
49. Yang, Y.: Consistency of cross validation for comparing regression procedures. Ann. Stat. **35**, 2450–2473 (2007)
50. Liao, H., Lynn, H.S., Li, S., Hsu, L., Peng, J., Wang, P.: Bootstrap inference for network construction with an application to a breast cancer microarray study. Ann. Appl. Stat. **7**, 391–417 (2013)
51. Waldron, L., Pintilie, M., Tsao, M.S., Shepherd, F., Huttenhower, C., Jurisica, I.: Optimized application of penalized regression methods to diverse genomic data. Bioinformatics **27**, 3399–3406 (2011)

52. Lê Cao, K.A., Boitard, S., Besse, P.: Sparse pls discriminant analysis: biologically relevant feature selection and graphical displays for multiclass problems. BMC Bioinform. **12**, 253 (2011)

53. Bunea, F., She, Y., Ombao, H., Gongvatana, A., Devlin, K., Cohen, R.: Penalized least squares regression methods and applications to neuroimaging. Neuroimage **55**, 1519–1527 (2011)

54. Rohart, F., Villa-Vialaneix, N., Paris, A., Laurent, B., SanCristobal, M.: Phenotypic prediction based on metabolomic data: lasso vs bolasso, primary data vs wavelet data. In: Proceedings of the 9th World Congress on Genetics Applied to Livestock Production (WCGALP), Leipzig, Germany (2010)

55. Avalos, M., Orriols, L., Pouyes, H., Grandvalet, Y., Thiessard, F., Lagarde, E.: Variable selection on large case-crossover data: application to a registry-based study of prescription drugs and road-traffic crashes. Pharmacoepidemiol. Drug Saf. **23**, 140–151 (2013). (Epub ahead of print)

56. Greenland, S.: Invited commentary: variable selection versus shrinkage in the control of multiple confounders. Am. J. Epidemiol. **167**, 523–529 (2008)

57. Walter, S., Tiemeier, H.: Variable selection: current practice in epidemiological studies. Eur. J. Epidemiol. **24**, 733–736 (2009)

58. Hurvich, C.M., Tsai, C.L.: The impact of model selection on inference in linear regression. Am. Stat. **44**, 214–217 (1990)

59. Breiman, L.: Heuristics of instability and stabilization in model selection. Ann. Stat. **24**, 2350–2383 (1996)

60. Austin, P.C.: Using the bootstrap to improve estimation and confidence intervals for regression coefficients selected using backwards variable elimination. Stat. Med. **27**, 3286–3300 (2008)

61. Wiegand, R.E.: Performance of using multiple stepwise algorithms for variable selection. Stat. Med. **29**, 1647–59 (2010)

62. Tibshirani, R.: The lasso method for variable selection in the Cox model. Stat. Med. **16**, 385–95 (1997)

63. Osborne, M.R., Presnell, B., Turlach, B.A.: On the lasso and its dual. J. Comput. Graph. Stat. **9**, 319–337 (2000)

64. Chatterjee, A., Lahiri, S.N.: Asymptotic properties of the residual bootstrap for lasso estimators. Proc. Am. Math. Soc. **138**, 4497–4509 (2010)

65. Park, M., Hastie, T.: $l_1$-regularization path algorithm for generalized linear models. J. Roy. Stat. Soc. Ser. B **69**, 659–677 (2007)

66. D'Angelo, G.M., Rao, D.C., Gu, C.C.: Combining least absolute shrinkage and selection operator (lasso) and principal-components analysis for detection of gene-gene interactions in genome-wide association studies. BMC Proc. **3**, S62 (2009)

67. Pötscher, B.: Confidence sets based on sparse estimators are necessarily large. Sankhya **71**, 1–18 (2009)

68. Pötscher, B., Schneider, U.: Confidence sets based on penalized maximum likelihood estimators in Gaussian regression. Electron. J. Stat. **4**, 334–360 (2010)

69. Farchione, D., Kabaila, P.: Variable-width confidence intervals in gaussian regression and penalized maximum likelihood estimators. Technical report, Department of Mathematics and Statistics, La Trobe University, Australia (2010)

70. Sperrin, M., Jaki, T.: Direct effects testing: a two-stage procedure to test for effect size and variable importance for correlated binary predictors and a binary response. Stat. Med. **29**, 2544–2556 (2010)

71. Goeman, J.: $l_1$ penalized estimation in the Cox proportional hazards model. Biometrical J. **52**, 70–84 (2010)

72. Sartori, S.: Penalized regression: Bootstrap confidence intervals and variable selection for high-dimensional data sets. PhD thesis, Raleigh, NC (2011)
73. Avalos, M., Duran-Adroher, N., Thiessard, F., Grandvalet, Y., Orriols, L., Lagarde, E.: Prescription-drug-related risk in driving comparing conventional and lasso shrinkage logistic regressions. Epidemiology **23**, 706–12 (2012)
74. Park, M., Casella, G.: The bayesian lasso. J. Am. Stat. Assoc. **103**, 681–686 (2008)
75. Hans, C.: Model uncertainty and variable selection in bayesian lasso regression. Stat. Comput. **20**, 221–229 (2010)
76. Sardy, S.: On the practice of rescaling covariates. Int. Stat. Rev. **76**, 285–297 (2008)
77. Belloni, A., Chernozhukov, V.: Least squares after model selection in high-dimensional sparse models. Bernoulli **19**, 521–547 (2013)
78. Bien, J., Taylor, J., Tibshirani, R.: A lasso for hierarchical interactions. Ann. Stat. **41**, 1111–1141 (2013)
79. Gertheiss, J., Tutz, G.: Sparse modeling of categorial explanatory variables. Ann. Appl. Stat. **4**, 2150–2180 (2010)
80. Avalos, M., Pouyes, H., Grandvalet, Y., Orriols, L., Lagarde, E.: Sparse conditional logistic regression for analyzing large-scale matched data from epidemiological studies: A simple implementation in r. Technical report, Bordeaux School of Public Health, University Bordeaux Segalen (2013) (Submitted)
81. Jörnsten, R., Abenius, T., Kling, T., Schmidt, L., Johansson, E., Nordling, T., Nordlander, B., Sander, C., Gennemark, P., Funa, K., Nilsson, B., Lindahl, L., Nelander, S.: Network modeling of the transcriptional effects of copy number aberrations in glioblastoma. Mol. Syst. Biol. **7**, Art. 486 (2011). doi:10.1038/msb.2011.17
82. Bunea, F.: Honest variable selection in linear and logistic regression models via $l_1$ and $l_1 + l_2$ penalization. Electron. J. Stat. **2**, 1153–1194 (2008)

# Piecewise Exponential Artificial Neural Networks (PEANN) for Modeling Hazard Function with Right Censored Data

Marco Fornili[1]([⊠]), Federico Ambrogi[1], Patrizia Boracchi[1], and Elia Biganzoli[1,2]

[1] Department of Clinical Sciences and Community Health,
University of Milan, Milan, Italy
{marco.fornili,federico.ambrogi,patrizia.boracchi,
elia.biganzoli}@unimi.it
[2] Unit of Medical Statistics, Biometry and Bioinformatics,
Fondazione IRCCS Istituto Nazionale Dei Tumori, Milan, Italy

**Abstract.** The hazard function plays an important role in the study of disease dynamics in survival analysis. Longer follow-up for various kinds of cancer, particularly breast cancer, has made it possible the observation of complex shapes of the hazard function of occurrence of metastasis and death. The identification of the correct hazard shape is important both for formulation and support of biological hypotheses on the mechanism underlying the disease.

In this paper we propose the use of a neural network to model the shape of the hazard function in time in dependence of covariates extending the piecewise exponential model. The use of neural networks accommodates a greater flexibility in the study of the hazard shape.

## 1 Introduction

Several analyses of survival data the role of prognostic factors is investigated independently of the shape of the baseline hazard function, which is commonly left undefined (Cox model). Several analyses of survival data are focused on prognostic factors independently of the shape of the baseline hazard function, which is commonly left undefined (Cox model). When the aim is to explore the time course of the disease to support or generate biological hypotheses, the shape of the hazard function provides relevant information as in the case of tumor dormancy in breast cancer [7].

When long follow-up is available, the shape of hazard function may be complex, for example with the presence of multiple peaks. Accordingly, sufficiently flexible techniques are required for an adequate estimation, for example kernel-like smoothing [15]. Other authors [10,18] adopted regression cubic splines with or without constraints of linearity on both tails based on a full likelihood parametric approach.

Further possibilities are the piecewise exponential and grouped-time models where the smoothing of the hazard is performed by splines [6]. In these two

© Springer International Publishing Switzerland 2014
E. Formenti et al. (Eds.): CIBB 2013, LNBI 8452, pp. 125–136, 2014.
DOI: 10.1007/978-3-319-09042-9_9

approaches time is subdivided in intervals, the main difference being that in each interval the piecewise exponential model directly estimates the hazard function, while the discrete time model estimates the conditional probability (discrete hazard). Artificial neural networks extending discrete-time model (PLANN [2]) have been proposed. In this paper we develop a neural network structure extending the piecewise exponential model (PEANN).

The complex structure of neural networks, with their well known approximating properties, allows in principle an extended flexibility in comparison to splines. In medical data neural networks could be useful when several variables are available and non-additive or non-linear effects are biologically plausible. In such situations a linear model where the structure of the linear predictor must be defined could be suboptimal.

In this paper an application to breast cancer data is presented where the shape of the hazard function could be hypothesized to depend on different kinds of events and/or covariates. Moreover, possible nonlinear effects for the continuous variables should be accounted for. Interactions could be expected too, as the prognostic role of a covariate may differ in time in dependence on both the other covariates and the type of event.

This paper is structured as follows: in Sect. 2 the technique is presented, in Sect. 3 the application is illustrated and the conclusions follow in Sect. 4.

## 2    Methods

In survival analysis, the distribution of the random variable time $U$ to a certain event is of interest. To model it, besides the density function $f(u)$ and the cumulative distribution function $F(u)$, the survivor function $S(u) = pr(T > u)$ and the hazard function

$$h(u) = \lim_{\Delta u \to 0^+} \frac{pr(u < T \leq u + \Delta u | T > u)}{\Delta u} \tag{1}$$

are often used. In particular, the latter is especially convenient when studying the dynamics of the event. The following relations link the aforementioned functions:

$$S(u) = exp\left(-\int_0^u h(s)ds\right) \tag{2}$$

$$f(u) = h(u)S(u). \tag{3}$$

Survival data are often right censored, i.e. the time to event is not observed and the information available is the last time of follow-up when the subject is event free. If $U_i$ and $C_i$ denote, respectively, the random variables time to event of interest and time to censorship for the $i$−th subject, we observe $t_i$, the realization of the variable $T_i = min(U_i, C_i)$, and the censoring index $\delta_i$, equal to 0 if $t_i$ is censored and 1 otherwise.

To estimate the hazard function we follow an approach based on the optimization of the likelihood function.

Under the hypothesis of non-informative censoring, the full likelihood for $N$ independent observations can be written as [11]:

$$L = \prod_{i=1}^{N} f(t_i; \mathbf{x}_i)^{\delta_i} S(t_i; \mathbf{x}_i)^{1-\delta_i} = \prod_{i=1}^{N} \frac{h(t_i; \mathbf{x}_i)^{\delta_i}}{exp\left(\int_0^{t_i} h(u; \mathbf{x}_i)du\right)}, \qquad (4)$$

where $\mathbf{x}_i = (x_{i1}, \ldots, x_{ip})$ denotes the $p$ covariates for subject $i$.

By introducing the assumption that the hazard function is constant within each of $J$ intervals $[a_{j-1}, a_j)$, $j = 1, \ldots, J$, where $a_0 = 0$ and $a_J = +\infty$, i.e. $h(t_i; \mathbf{x}_i) = \lambda_j(\mathbf{x}_i)$ for $a_{j-1} \leq t_i < a_j$, we can write [1]:

$$L = \prod_{i=1}^{N} \frac{\prod_{j=1}^{J_i} \lambda_j(\mathbf{x}_i)^{\delta_{ij}}}{exp\left(\sum_{j=1}^{J_i} \lambda_j(\mathbf{x}_i)\tau_{ij}\right)} = \frac{1}{\prod_{i=1}^{N} \prod_{j=1}^{J_i} \tau_{ij}^{\delta_{ij}}} \prod_{i=1}^{N} \prod_{j=1}^{J_i} \frac{(\lambda_j(\mathbf{x}_i)\tau_{ij})^{\delta_{ij}}}{\delta_{ij}! exp\left(\lambda_j(\mathbf{x}_i)\tau_{ij}\right)},$$

$$(5)$$

where

$$\delta_{ij} = \begin{cases} 1 \text{ if subject } i \text{ experiments the event in the } j-th \text{ interval} \\ 0 \text{ otherwise} \end{cases}$$

$$\tau_{ij} = \begin{cases} a_j - a_{j-1} \text{ if } a_j \leq t_i \\ t_i - a_{j-1} \text{ if } a_{j-1} < t_i \leq a_j \\ 0 \qquad \text{ if } t_i \leq a_{j-1} \end{cases}$$

and $J_i$ indicates the last interval in which subject $i$ is observed. Represents, apart from a constant factor, the likelihood corresponding to observations $\delta_{ij}$ from independent Poisson variables with means $\mu_{ij} = \lambda_j(\mathbf{x}_i)\tau_{ij}$.

We model the hazard rates $\lambda_j$'s as a function of time as well as of covariates: $\lambda_j(\mathbf{x}_i) = h(\xi_j; \mathbf{x}_i)$, where $\xi_j$ is the midpoint of the $j$-th time interval and $h(\xi; \mathbf{x})$ is a smooth function.

Usually Generalized Linear Models (GLM) with Poisson error are adopted for inference. $h(\xi; \mathbf{x})$ is then a monotonic transformation of a linear predictor in which time and covariates are modelled by splines. We propose to model $h(\xi; \mathbf{x})$ by artificial neural networks [3,4,12,17], which represent a flexible tool successfully adopted in statistics to flexibly model the dependence on covariates of continuous, dichotomous and polytomous variables. Theoretical results on the approximation capabilities of neural networks have been obtained; for example it has been proved that every continuous function can be approximated uniformly on a compact set by a neural network with linear output activation and logistic hidden units activation. Starting from the recent proposals of neural networks with Poisson responses [8,14], we extended the methods to account for the presence of censored survival data.

In this paper the hazard function is modeled by means of the following feed-forward artificial neural network:

$$h(t; \mathbf{x}) = exp\left(\beta_0^{(2)} + \sum_{k=1}^{H} logis\left(\beta_{0k}^{(1)} + \sum_{l=1}^{p} x_l \beta_{lk}^{(1)} + t\beta_{p+1,k}^{(1)}\right)\beta_k^{(2)}\right), \qquad (6)$$

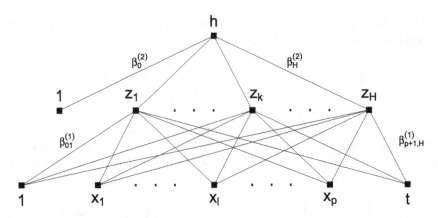

**Fig. 1.** Scheme of a multilayer perceptron with the layer of inputs $x_1, \ldots, x_p, t$, the layer of hidden units $z_1, \ldots, z_H$ and the output layer, here represented by the unique response h. A constant unit set to 1 feeds every other non-input unit for clarity it is indicated twice.

where *exp* and *logis* stand for the exponential and the logistic function respectively.

This net (Fig. 1) presents three layers of units: the inputs $x_l$ and $t$, the hidden units $z_k = logis\left(\beta_{0k}^{(1)} + \sum_{l=1}^{p} x_l \beta_{lk}^{(1)} + t\beta_{p+1,k}^{(1)}\right)$ and the single output $h$. Moreover a constant unit set to 1 "feeds" every non-input unit (for clarity of representation in the Fig. 1 it has been indicated twice).

To estimate the coefficients $\beta$ we minimize the error function

$$E = -log\, L + \alpha \left( \sum_{k=0}^{H} \sum_{l=0}^{p+1} \beta_{lk}^{(1)2} + \sum_{k=0}^{H} \beta_k^{(2)2} \right), \tag{7}$$

where the first term is the negative of the logarithm of the likelihood and the second term is a quadratic penalty [17]. The effect of the penalization is that of increasing the performance of the optimization routines and of preventing overfitting. The decay parameter $\alpha$ regulates the trade-off between smoothness and fitting to the data.

The error function can be explicitly written in terms of the augmented data $(\xi_j, \mathbf{x}_i, \tau_{ij}, \delta_{ij})$ as

$$E = -\sum_{i=1}^{N} \sum_{j=1}^{J_i} [\delta_{ij} \log h(\xi_j; \mathbf{x}_i) - h(\xi_j; \mathbf{x}_i)\tau_{ij}] + \alpha \left( \sum_{k=0}^{H} \sum_{l=0}^{p+1} \beta_{lk}^{(1)2} + \sum_{k=0}^{H} \beta_k^{(2)2} \right).$$

$$\tag{8}$$

To assure that in the penalty the coefficients from input to hidden units and those from hidden to output units are of comparable size, the input values are properly scaled to lie between about -1 and 1 and the initial values of the coefficients furnished to the optimization routine are sampled from a uniform distribution on the interval (-0.7, 0.7).

For the optimization we used the quasi-Newtonian algorithm BFGS. It exploit the gradient of the error function, which is given by:

$$\frac{\partial E}{\partial \beta_0^{(2)}} = -\sum_{i=1}^{N}\sum_{j=1}^{J_i} d_{ij} + 2\alpha\beta_0^{(2)} \tag{9}$$

$$\frac{\partial E}{\partial \beta_k^{(2)}} = -\sum_{i=1}^{N}\sum_{j=1}^{J_i} d_{ij}z_{ik} + 2\alpha\beta_k^{(2)} \tag{10}$$

$$\frac{\partial E}{\partial \beta_{0k}^{(1)}} = -\sum_{i=1}^{N}\sum_{j=1}^{J_i} d_{ij}\beta_k^{(2)} z_{ik}\left(1 - z_{ik}\right) + 2\alpha\beta_{0k}^{(1)} \tag{11}$$

$$\frac{\partial E}{\partial \beta_{lk}^{(1)}} = -\sum_{i=1}^{N}\sum_{j=1}^{J_i} d_{ij}\beta_k^{(2)} z_{ik}\left(1 - z_{ik}\right)x_{il} + 2\alpha\beta_{lk}^{(1)}, \tag{12}$$

where $d_{ij} = \delta_{ij} - h(\xi_j; \mathbf{x}_i)\tau_{ij}$.

The flexibility of the neural networks depends on both the number of hidden units and the decay parameter. One strategy to choose them is to recur to cross-validation. We adopted a ten-fold cross validation, by splitting the subjects into ten subsets $S^{(m)}$, $m = 1, \ldots, 10$ of about the same size. The criterion to be minimized is then

$$CV = \frac{1}{\sum_{i=1}^{N} J_i} \cdot \sum_{i=1}^{N}\sum_{j=1}^{J_i} \left[ \delta_{ij} \log \hat{h}^{(-m(i))}(\xi_j; \mathbf{x}_i) - \hat{h}^{(-m(i))}(\xi_j; \mathbf{x}_i)\tau_{ij} \right], \tag{13}$$

where $\hat{h}^{(-m(i))}$ is the hazard estimate obtained by averaging on five fits for various values of the number of hidden units and of the decay parameter. These fits were performed by excluding individuals which belong to the same subset $S^{(m(i))}$ as subject $i$.

As cross validation is computationally intensive, we chose values of the decay parameter in the explorative range of $10^{-3} - 10^{-1}$, which was suggested in [17] based on Bayesian considerations. Alternatively, the value of the decay parameter can be selected via information criteria. Among the latter, one of the most used, NIC, has been proved [16] to be equivalent for large samples to the leave-one-out cross validation. Nevertheless NIC may be less reliable when many local minima are present.

In performing both cross validation and estimation, the averaging over multiple fits obtained from different initial values for the parameters has been advocated [19].

When different kinds of event are of concern, the hazard for each event is named cause-specific hazard (CSH) and a covariate indicating the type of event has to be included in the covariate vector. The likelihood function (Eq. 5) can be generalized as reported in detail in [6].

We used R [13], version 3.0.1, for all computations, in particular the package *Epi* for data augmentation and function *optim* for optimization.

## 3    Breast Cancer Survival Study

As an application we consider a dataset of 2233 breast cancer patients hospitalized at the Istituto Nazionale dei Tumori of Milan between 1970 and 1987 [20]. All patients underwent conservative surgery and axillary lymph node dissection followed by radiotherapy. Information about age (in years), tumor size (in cm) and number of metastatic axillary nodes was recorded. Histological type (categorized as: extensive intraductal component; infiltrating intraductal or infiltrating lobular component; other histologies) and tumor site (categorized as: external quadrant; internal or central quadrant) are also available. With a median follow-up of 8.5 years, 151 intrabreast tumor recurrences (IBTRs), 110 contralateral breast carcinomas, 414 distant metastases (DMs) and 69 primary non-mammary malignant tumors were recorded as first neoplastic events. 27 patients died without any evidence of breast cancer recurrence. The interval until the appearance of neoplastic events other than IBTR and DM was censored.

For the dataset under study, to select the structure of the network, a ten-fold cross validation was applied considering 3, 6 and 9 hidden units and penalty coefficients 0.001, 0.01 and 0.1. The values of the CV criterion (Table 1) show a better performance in correspondence to a choice of 6 hidden units and decay parameter equal to 0.01, that we therefore used.

We then considered a neural network with time, age, tumor size and number of metastatic lymph nodes as continuous variables, and histological type, tumor site and type of event as categorical variables, using the average over ten fits obtained from different initial choice of parameters. Due to the possible presence of non-additive effects, the investigated shapes might vary among different values of the conditioning covariates. A detailed exploration was accomplished by multipanel conditioning plots to investigate the joint effects of couples of continuous covariates on the shape of CSHs. For each variable, three selected values were considered. With regards to age, 30, 50 and 70 years were chosen, for tumour size 0.5, 1.4, 2.5 cm, and, for the number of metastatic axillary nodes, 0, 3 and 10. A different behavior of the three variables on IBTR and DM is observed. The hazard function trend for the two failure causes show different patterns. As regards IBTR CSH, a strong impact of a young age and largest tumor size

**Table 1.** Ten-fold cross-validation obtained by averaging over five fits for various values of the number of hidden units (H) and of the decay parameter.

| Decay | H = 3 | H = 6 | H = 9 |
|-------|-------|-------|-------|
| 0.001 | 0.2576 | 0.2727 | 4.3013 |
| 0.01 | 0.2570 | 0.2562 | 0.3409 |
| 0.1 | 0.2568 | 0.2564 | 0.2565 |

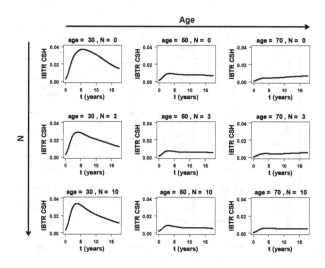

**Fig. 2.** Multipanel conditioning plots of cause-specific hazards for intrabreast tumor recurrence (IBTR CSH) for age = 30, 50 and 70 years and number of positive lymph nodes (N) = 0, 3 and 10. Tumor size is fixed at its median values (1.4 cm) while categorical covariates are fixed at their modal values (tumor site = external, histology = infiltrating intraductal or infiltrating lobular component).

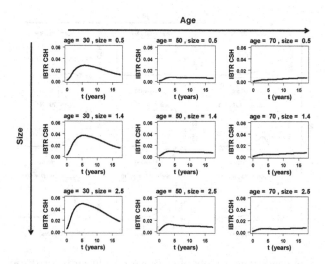

**Fig. 3.** Multipanel conditioning plots of cause-specific hazards for intrabreast tumor recurrence (IBTR CSH) for age = 30, 50 and 70 years and tumor size = 0.5, 1.4 and 2.5 years. Number of positive nodes is fixed at its median values (0) while categorical covariates are fixed at their modal values (tumor site = external, histology = infiltrating intraductal or infiltrating lobular component).

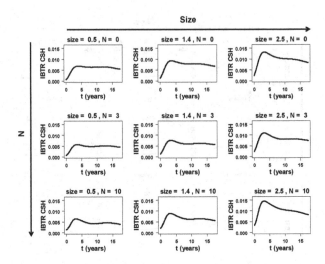

**Fig. 4.** Multipanel conditioning plots of cause-specific hazards for intrabreast tumor recurrence (IBTR CSH) for tumor size = 0.5, 1.4 and 2.5 years and number of positive lymph nodes (N) = 0, 3 and 10. Age is fixed at its median values (50 years) while categorical covariates are fixed at their modal values (tumor site = external, histology = infiltrating intraductal or infiltrating lobular component).

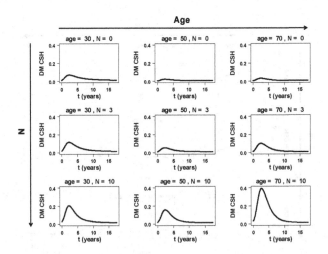

**Fig. 5.** Multipanel conditioning plots of cause-specific hazards for distant metastasis (DM CSH) for age = 30, 50 and 70 years and number of positive lymph nodes (N) = 0, 3 and 10. Tumor size is fixed at its median values (1.4 cm) while categorical covariates are fixed at their modal values (tumor site = external, histology = infiltrating intraductal or infiltrating lobular component).

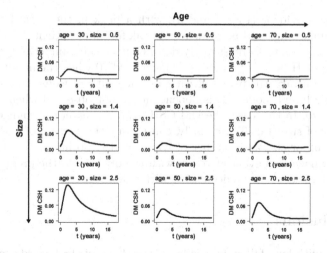

**Fig. 6.** Multipanel conditioning plots of cause-specific hazards for distant metastasis (DM CSH) for age = 30, 50 and 70 years and tumor size = 0.5, 1.4 and 2.5 years. Number of positive nodes is fixed at its median values (0) while categorical covariates are fixed at their modal values (tumor site = external, histology = infiltrating intraductal or infiltrating lobular component).

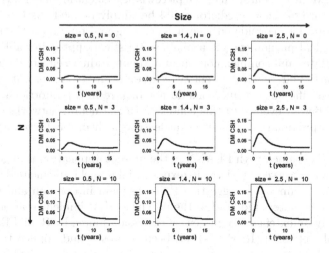

**Fig. 7.** Multipanel conditioning plots of cause-specific hazards for distant metastasis (DM CSH) for tumor size = 0.5, 1.4 and 2.5 years and number of positive lymph nodes (N) = 0, 3 and 10. Age is fixed at its median values (50 years) while categorical covariates are fixed at their modal values (tumor site = external, histology = infiltrating intraductal or infiltrating lobular component).

is apparent while the hazard shape in dependence of the number of positive lymph nodes is approximately constant (Figs. 2, 3 and 4). No pairwise interaction between age, size and N is evident. No time-dependent effect is observed for tumor size: a peak of the hazard is shown at about 2 years. A time-dependent effect of age is evident (Figs. 2 and 3), with a peak of the hazard at about 5 years for youngest age and no evidence of peak for older patients. For DM CSH, the effect of age seems to be strongly dependent on the number of lymph nodes (Fig. 5). The CSH peak is greater for women aged 70 when the number of nodes is 10; on the contrary, the peak is greater for the youngest women when there are no metastatic lymph nodes. This dependence is not observed when age is conditioned to size. In this case, the DM CSH peak is greater for youngest women for all values of size (Fig. 6). Similarly, considering the pattern of DM CSH as a function of tumor size and number of lymph nodes, no interaction is apparent (Fig. 7). The hazard of event increases with tumor size and the prognostic effect of the covariate seems to be decreasing with time.

## 4    Conclusions

For diseases with long follow-up available, the shape of the hazard function may be of great clinical interest to support clinical-biological hypotheses or to plan follow-up visits. In the past, a limited number of covariates, preferably on a categorical scale, were recorded. Previous knowledge about the effect of covariates was available, so that the main task was smoothing the effect of time. For this purpose, GLM implementation of the piecewise exponential model was sufficient, as the structure of linear predictor could be simply defined. As the number of prognostic variables considered is ever increasing and only a partial knowledge on the biological phenomenon is available, statistical approaches able to model jointly covariates and complex non-linear and non-additive effects should be preferred. In particular, when different types of event are of concern, it is expected that cause-specific hazard functions do not behave in a proportional way with respect to different events. In this context artificial neural networks, in view of their good approximation properties, at least at a theoretical level, are a useful tool.

The trend obtained with PEANN in the present application is in good agreement with that observed by piecewise exponential model implemented with GLM and regression splines [6], suggesting that the main interaction and non-linear effects were adequately modeled in the previous GLM implementation.

Previously an extension of piecewise regression model by Radial Basis Functions Neural Networks (RBFNNs) has been proposed and applied to the same data set [5]. An attractive feature of this approach is that it can be implemented with standard statistical software for GLM. In fact, localized symmetric basis functions are directly included in the model linear predictor. The estimates of IBTR and DM cause-specific hazards by RBFNN appear to be generally consistent with those obtained by PEANN, but show more complex patterns in the late follow-up time. Nevertheless special attention should be paid to the examination of right tail behavior of the hazard functions since the RBFNNs estimates, like those obtained with non-parametric procedures, may fluctuate. A possible drawback of RBFNNs is the choice of the centers and widths of the radial basis functions. In particular heuristic approaches have been proposed allowing fast modeling implementation but not necessarily providing an optimal performance [3]. PEANN, like the other feed forward neural networks, does not present these problems.

As concerns the selection of the decay parameter and the number of hidden units, a simulation study would be advisable to compare the performance of cross validation and information criteria.

In conclusion, artificial neural networks appear to be useful in the intermediate steps of the model building, both for exploratory evaluation of non-linear and non-additive effects and for checking model assumptions.

# References

1. Aitkin, M., Francis, B., Hinde, J., Darnell, R.: Statistical Modelling in R. Oxford University Press, Oxford (2009)
2. Biganzoli, E., Boracchi, P., Marubini, E.: A general framework for neural network models on censored survival data. Neural Netw. **37**, 119–130 (2002)
3. Bishop, C.M.: Neural Networks for Pattern Recognition. Springer, New York (1995)
4. Bishop, C.M.: Pattern Recognition and Machine Learning. Springer, New York (2006)
5. Boracchi, P., Biganzoli, E., Marubini, E.: Modelling cause-specific hazards with radial basis function artificial neural networks: application to 2233 breast cancer patients. Statist. Med. **20**, 3677–3694 (2001)
6. Boracchi, P., Biganzoli, E., Marubini, E.: Joint modelling of cause-specific hazard functions with cubic splines: an application to a large series of breast cancer patients. Comput. Stat. Data An. **42**, 243–262 (2003)
7. Demicheli, R., Retsky, M.W., Baum, M.: Tumor dormancy and surgery-driven interruption of dormancy in breast cancer: learning from failures. Nat. Clin. Pract. Oncol. **4**, 699–710 (2007)
8. Fallah, N., Gu, H., Mohammad, K., Seyyedsalehi, S.A., Nourijelyani, K., Eshraghian, M.R.: Nonlinear Poisson regression using neural networks: a simulation study. Neural Comput. Appl. **18**, 939–943 (2009)
9. Fu, K.K., Phillips, T.L., Silverberg, I.J., Jacobs, C., Goffinet, D.R., Chun, C., Friedman, M.A., Kohler, M., McWhirter, K., Carter, S.K.: Combined radiotherapy and chemotherapy with bleomycin and methotrexate for advanced inoperable head and neck cancer: update of a Northern California Oncology Group randomized trial. J. Clin. Oncol. **5**, 1410–1418 (1987)
10. Herndon, J.E., Harrell, F.E.: The restricted cubic spline hazard model. Comm. Statist. Theory Meth. **19**, 639–663 (1990)
11. Marubini, E., Valsecchi, M.G.: Analysing Survival Data from Clinical Trials and Observational Studies. Wiley, Chichester (2004)
12. Nabney, I.T.: NETLAB, Algorithms for Pattern Recognition. Springer, New York (2002)
13. R Core Team: R: A language and environment for statistical computing. R Foundation for Statistical Computing, Vienna, Austria (2013). http://www.R-project.org/
14. Rajaram, S., Graepel, T., Herbrich, R.: Poisson-networks: a model for structured point processes. In: Proceedings of the Tenth International Workshop on Artificial Intelligence and Statistics (2005)

15. Ramlau-Hansen, H.: Smoothing counting process intensities by means of kernel functions. Ann. Stat. **11**, 453–466 (1983)
16. Ripley, B.D.: Statistical ideas for selecting network architectures. In: Kappen, B., Gielen, S. (eds.) Neural Networks: Artificial Intelligence and Industrial Applications, pp. 183–190. Springer, London (1995)
17. Ripley, B.D.: Pattern Recognition and Neural Networks. Cambridge University Press, Cambridge (1996)
18. Rosenberg, P.S.: Hazard function estimation using B-splines. Biometrics **51**, 874–887 (1995)
19. Venables, W.N., Ripley, B.D.: Modern Applied Statistics with S. Springer, New York (2002)
20. Veronesi, U., Marubini, E., DelVecchio, M., Manzari, A., Andreola, S., Greco, M., Luini, A., Merson, M., Saccozzi, R., Rilke, F., Salvadori, B.: Local recurrences and distant metastases after conservative breast cancer treatments: partly independent events. J. Natl Cancer Inst. **87**, 19–27 (1995)

# Writing Generation Model for Health Care Neuromuscular System Investigation

D. Impedovo[1], G. Pirlo[1], F.M. Mangini[1], D. Barbuzzi[1], A. Rollo[1],
A. Balestrucci[1], S. Impedovo[1(✉)], L. Sarcinella[2], C. O'Reilly[3],
and R. Plamondon[3]

[1] Computer Science Department, Bari University, Via Orabona 4, Bari, Italy
sebastiano.impedovo@uniba.it
[2] Rete Puglia Centre, Via G. Petroni 15/F.1, Bari, Italy
[3] École Polytecnique de Montréal, Montreal, Canada

**Abstract.** In this paper the use of handwriting for health investigation is addressed. For the purpose, the paper first presents the Delta-Log and Sigma-Log models to investigate on the handwriting generation processes carried out by the neuromuscular system. Successively, a computational system for handwriting analysis is presented and some considerations are exploited about the use of the model to investigate insurgence and monitoring of some neuromuscular diseases. The experimental results show the validity of the proposed approach and highlight some directions for further research.

**Keywords:** Neuromuscular disease investigation · Handwriting analysis · Neuromuscular transfer function

## 1 Introduction

In the developed societies, neuromuscular diseases are generally the leading cause of greater severe adult disability. In the modern society, these disabilities generate health-care expenditures due to hospitalization, rehabilitation, long-term cares and so on. Therefore a severe investigation is strongly required to know better brain activities in connection with the insurgence of some illnesses.

The investigation of the upper limb human movements like those supporting handwriting generation can become one of the strategies to neuromuscular illness assessment. In this paper, pen-strokes modelling in handwriting generation is discussed and its use for diagnostic investigation is addressed. The investigations of signal degeneration related to Alzheimer disease growth is here developed, starting from the early stage of insurgence, by taking into account the handwriting actions.

To investigate in depth the phenomena, in this paper a short survey of the kinematic theory is firstly presented by considering both delta and sigma lognormal models and the different algorithms for the extraction of parameters describing handwriting stroke trajectories [1]. Then the neuromuscular task selected for diagnostic system is described by pointing out the assessing of its performances. Specifically, many design questions associated to data acquisition, motion deterministic modelling, and decision making are presented.

© Springer International Publishing Switzerland 2014
E. Formenti et al. (Eds.): CIBB 2013, LNBI 8452, pp. 137–148, 2014.
DOI: 10.1007/978-3-319-09042-9_10

The results can be then used for developing a statistical model leading to a diagnostic protocol of assessment of the brain illness and monitoring of the cerebral disease.

## 2  Neuromuscular System Function Transfer – the Delta-Log Model

Based on a kinematic theory of rapid movement generation [2], the proposed model allows the description of individual strokes in terms of a delta lognormal equation. It predicts the very basic speed/accuracy tradeoffs observed in target-directed movements.

Several bottom-up approaches have been used by those interested in the analysis and synthesis of the low-level neuromuscular process, obviously by controlling agonist and antagonist muscular movements. For these perceptive motor strategies involved in the generation and perception of handwriting, two criteria have been met:

1.  A model should be realistic enough to reproduce specific pen tip trajectories perfectly;
2.  Its descriptive power should be such that it provides consistent explanations of the basic properties of single stroke.

**Delta Lognormal Model**

A general way to look at the impulse response of a specific controller, say the velocity module controller, is to consider the overall sets of neural and muscle networks involved in the production of a single stroke as a synergetic linear system producing a curvilinear velocity profile from an impulse command of amplitude D occurring at the time t0. So the curvilinear velocity profile directly reflects the impulse response H(t-t0) of neuromuscular synergy [1, 3].

The mathematical description of this impulse response can be specified by considering each controller as composed of two systems that represent the sets of neural and muscular networks involved in the generation of the agonist and antagonist activities resulting in a specific movement.

In the mathematical description, agonist and antagonist impulses are taken into account. It is expected that the impulse response of a system under the coupling hypothesis will converge toward a log-normal curve. The output of the module or the direction controller will be described by the weighted difference of two lognormal, called a delta lognormal equation [4].

In this context, the control of the velocity module can now be seen as resulting from the simultaneous activation (at time $t = t_0$) of a controller made up of two antagonist neuromuscular systems, with a command of amplitude $D_1$ and $D_2$ respectively. Both systems react to their specific commands with an impulse response described by a lognormal function, whose parameters $\mu_1$, $\sigma_1^2$ and $\mu_2$, $\sigma_2^2$ characterize the time delay and response time of each process.

$$v_\sigma = \frac{D_1}{\sigma_1\sqrt{2\pi}(t-t_0)}\exp\left[\frac{-(\ln(t-t_0)-\mu_1)^2}{2\sigma_1^2}\right]$$
$$-\frac{D_2}{\sigma_2\sqrt{2\pi}(t-t_0)}\exp\left[\frac{-(\ln(t-t_0)-\mu_2)^2}{2\sigma_2^2}\right] \tag{1}$$

In a similar fashion, for rotation movements like wrist flexion and extension, it is predicted that the control of the angular velocity will also obey a specific delta lognormal equation with respect to command amplitude $D_1$ and $D_2$. (I am not sure if I understood what you meant here but the sentence as it was written was not clear. If my correction is not what you meant feel free to make a better modification) Figure 1 shows the Delta lognormal profile obtained using these parameters:

- $D_1 = 2.35$ cm;
- $D_2 = 0.94$ cm;
- $t_0 = 0.348$ s;
- $\mu_1 = -1.61$;
- $\mu_2 = -1.45$;
- $\sigma_1^2 = 0.0134$;
- $\sigma_2^2 = 0.00497$;
- mean square error: $0.0728$ cm$^2$/s$^2$.

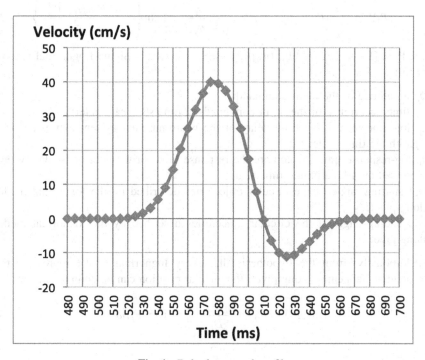

**Fig. 1.** Delta lognormal profile

**Model Testing: First Criterion**

The delta lognormal equation has been proven to be the most powerful in reconstructing experimental data and movement generation [4].

**Model Testing: Second Criterion**

It has been shown tendency of human subjects to produce strokes of about the same duration and the same relative spatial precision to insure legibility [2]. These evidences demonstrate that the delta lognormal model is certainly among the most promising candidate for sensor motor control of handwriting generation and perception.

## 3 Neuromuscular System Function Transfer – the Sigma-Log Model

The kinematic theory of rapid human movement, relies on the Sigma-Lognormal model to represent the information of both the motor commands and timing properties of the neuromuscular system involved in the production of complex movements like word writing [5, 6].

The Sigma-Lognormal model considers the resulting speed of a single stroke j as having a lognormal shape $\Lambda$ scaled by a command parameter (D) and time-shifted by the time occurrence of the command (t0).

$$\left|\vec{v}_j(t; P_j)\right| = D_j \Lambda(t - t_{0j}; \mu_j, \sigma_j^2) = \frac{D_j}{\sigma(t - t_{0j})\sqrt{2\pi}} \exp\left\{ \frac{\left[\ln(t - t_{0j}) - \mu_j\right]^2}{-2\sigma_j^2} \right\} \tag{2}$$

where $P_j = [D_j, t_{0j}, \mu_j, \sigma_j, \Theta_{sj}, \Theta_{ej}]$ represents the sets of Sigma-Lognormal parameters [5]:

- $D_j$: amplitude of the input commands;
- $t_{0j}$: time occurrence of the input commands, a time-shift parameter;
- $\mu_j$: log-time delays, the time delay of the neuromuscular system expressed on a logarithmic time scale;
- $\sigma_j$: log-response times, which are the response times of the neuromuscular system expressed on a logarithmic time scale;
- $\Theta_{sj}$: starting angle of the circular trajectories described by the lognormal model along pivot;
- $\Theta_{ej}$: ending angle of the circular trajectories described by the lognormal model along pivot.

Additionally, from the hypothesis that every lognormal stroke represents the movement as happening along pivot, the angular position can be computed as [5, 7]:

$$\phi_j(t; P_j) = \theta_{sj} + \frac{\theta_{ej} - \theta_{sj}}{D_j} \int_0^t \left|\vec{v}(\tau; P_j)\right| d\tau \tag{3}$$

In this context, a written word can be seen as the output of a generator that produces a set of individual strokes superimposed in time. The resulting complex trajectories can be modelled as a vector summation of lognormal distributions (being NLN the total number of lognormal curves in which the handwritten trace is decomposed):

$$\overrightarrow{v}(t) = \sum \Lambda(t) = \sum_{j=1}^{N_{LN}} \overrightarrow{v}_j(t; P_j) \qquad (4)$$

The velocity components in the Cartesian space can be calculated from the tangential speed as [7–9]:

$$\overrightarrow{v_x}(t) = \sum_{j=1}^{N_{LN}} \left| \overrightarrow{v_j}(t; P_j) \right| \cos(\phi_j(t; P_j))$$

$$\overrightarrow{v_y}(t) = \sum_{j=1}^{N_{LN}} \left| \overrightarrow{v_j}(t; P_j) \right| \sin(\phi_j(t; P_j)) \qquad (5)$$

The reconstruction error of a velocity profile using Sigma-Lognormal parameters can be evaluated by comparing the SNR between the reconstructed specimen and the original one:

$$10 \log \left( \frac{\int_{t_s}^{t_e} \left[ v_{xo}^2(t) + v_{yo}^2(t) \right] dt}{\int_{t_s}^{t_e} \left[ \left( v_{xo}(t) - v_{xa}(t) \right)^2 + \left( v_{yo}(t) - v_{ya}(t) \right)^2 \right] dt} \right) \qquad (6)$$

where $t_s$ and $t_e$ are respectively the starting and the ending times of the written word.

As it is well known enormous progress in signal processing has been done by using Fourier Analysis, in fact in this approach not only the time domain has been investigated but also the spectral analysis of trace has been started [10].

## 4   Experimental Setup for Writing Generation Model

Firstly, some clear effects on signature shape modification are shown in Figs. 2 and 3, in which it is visible the changing in signature shape by passing through the different period of the live for a person that approaching to the 90-th years of her live has been affect by the Alzheimer disease.

**Fig. 2.** Signature produced in year 1997, when the person was healthy

**Fig. 3.** Signature produced in year 2010, when the person was affected by Alzheimer disease

In order to investigate in depth the degenerative process, the software used for handwriting acquisition and parameter generation of the sigma-log model is considered. In particular the software determines the number of sigma-log functions and the parameter set of each function: ($t_0$, D, $\mu$, $\sigma$, $\theta_s$, $\theta_e$). The set of parameters conveys several information that are here used for analysis of the neuromuscular system in order to detect the illness progression of the neuromuscular disease.

In particular, concerning neuromuscular disorders, handwriting dynamics can be used for both early diagnosis of neuromuscular illness and monitoring evolution of the illness over time also to evaluate the individual effect of a patients treatment.

In order to start experiments and to tune the system, some initial evaluation have been done by using the ISUNIBA database developed for the purpose of this research at the Computer Science Department of Bari University, Italy. 50 patients affected by Alzheimer disease have been enrolled for the data base realization. Each patient supplied 10 of his/her handwriting words in two different sessions(5 for each session) collected in two separate sessions that were approximately one month a part. In order to monitoring the disease progression it has been programmed the repetition of data acquisition monthly. Figure 4 shows the acquisition device used for the experiments: the WACOM INTUOS-5 peripheral device.

**Fig. 4.** An example of using WACOM INTUOS-5

**Software System for Delta-Log Parameter Extraction**

The system that has been used for parameter extraction is named ScriptStudio, it has been developed at the Corporation de l'Ecole Polytechnique de Montreal, Canada [5].

For the aims of this paper two modules of the software have been considered:

- The first module has the objective to acquire handwritten words;
- The second module has the objective to generate parameters.

Each handwritten word is modelled in a matrix, in which each column has the following information:

- Column 1: time of starting of each point: $t_0$;
- Columns 2, 3, 4: position of each point in X, Y and Z axe, respectively;
- Column 5: pressure;
- Column 6: azimuth;
- Column 7: elevation;
- Column 8: tangential speed.

These information are modeled in a.hws file. that also includes information about file format, sampling rate and numbers of elements. An example of a.hws file is provided in the following.

```
TypeFichier: HWS
NombrePoints: 1202
FrequenceEchantillonage(Hz): 200.00
1.360000 0.503890 0.000000 0.000000 0.000000
0.000000  0.000000 147.402571
.......
7.365000 140.950904 -79.755773 0.000000 0.000000
0.000000  0.000000 13.732579
```

Each word is stored, stroke by stroke, in a .ana file. Specifically each element of the stroke matrix includes:

- Column 1: stroke initial time;
- Column 2: stroke amplitude;
- Columns 3, 4: log time delay and log response time, respectively;
- Columns 5, 6: initial and final angle of the stroke, respectively;
- The last three columns have to be unconsidered.

In the following there is the .ana file of an hws example.

```
TypeFichier: ANA
Version: 3
NombreLognormales: 17
1.820250 88.210019 -0.961191 0.375332 -1.128164 -
1.678627 0.000 0.000 1
.....
1.716305 15.691989 -1.599541 0.384121 -1.204084 -
1.270359 0.000 0.000 1.
```

The same analysis has been developed to investigate modifications in handwriting words due to neuromuscular diseases. In the following, the results of some pattern analysis related to patients are reported.

In order to investigate in depth the relationship between handwriting and Alzheimer disease, a detailed investigation on 50 patients has been developed by acquiring some handwriting basic words like "mamma". In fact the word "mamma" is one of the first learned and written and remain one of the last used before to died.

In order to investigate the relation between disease and writing process also the Fourier analysis of writing trace has been developed and the spectra for three handwriting words, the first produced by an healthy person, the second by a patient in the early stage of Alzheimer disease and the last related to an advanced stadium of disease are reported in Figs. 5, 6 and 7, respectively.

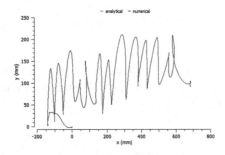

**Fig. 5.** Example of handwritten word "mamma" made by an healthy person

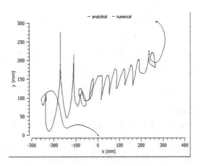

**Fig. 6.** Example of handwritten word "mamma" made by a patient in early stage of illness

The .ana profiles in Figs. 8, 9 and 10 show respectively the velocity profiles obtained by the ANA files.

By looking the speed profile along the process, it can be observed that in healthy person the maxima speed value are almost regular in height (Fig. 8), instead these regularity is strongly reduced at beginning of the disease (Fig. 9) and completely lost in the advance stadium of the disease (Fig. 10).

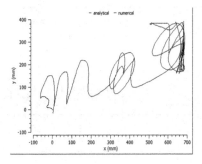

**Fig. 7.** Example of handwritten word "mamma" made by a patient in advanced stadium of illness

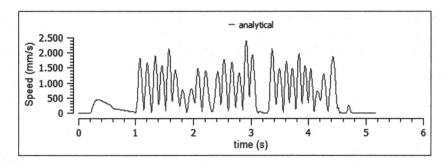

**Fig. 8.** Velocity profile for word in Fig. 5

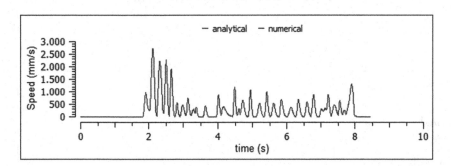

**Fig. 9.** Velocity profile for word in Fig. 6

Furthermore it must be noted that the time duration pass from 3.5 s up to 6 s to reach the duration of 10 s as can be seen comparing Figs. 8, 9 and 10.

It must be also noted that the starting delay pass from 1 s. to about 2 s, to reach 4.5 s for people heavily affected by the disease.

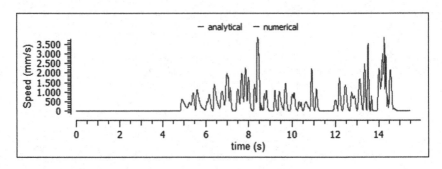

**Fig. 10.** Velocity profile for word in Fig. 7

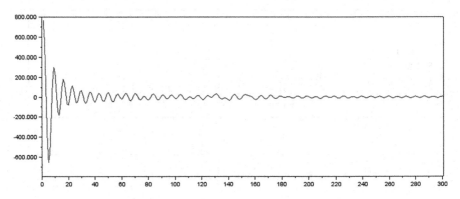

**Fig. 11.** Spectral analysis for word in Fig. 5

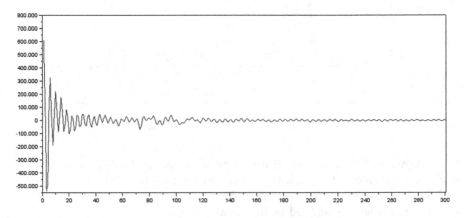

**Fig. 12.** Spectral analysis for word in Fig. 6

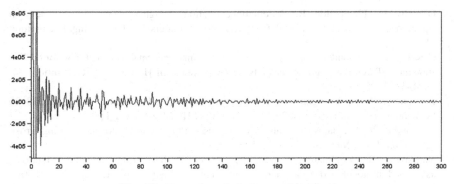

**Fig. 13.** Spectral analysis for word in Fig. 7

Furthermore by looking more in depth the velocity profiles it can be seen that for healthy person rarely the speed becomes zero instead for injured patients the zero speed is reached many times.

It must be also noted that the duration of the zero profile is enlarged for injured persons.

The spectra profiles reported in Figs. 11, 12 and 13, clearly shown the Alzheimer effect when the number of maxima along the spectra are considered in the frequency band 1–40 Hz. In fact, the number of peaks increase from 6 in healthy person to 10 at beginning of disease to reach 15 in advanced stadium of the illness.

## 5  Conclusion

In this paper it has been introduced the investigation of handwriting words in order to detect some parameters for Alzheimer diagnosis and its disease process monitoring. Both parameters in time domain and in frequency domain have been shown.

**Acknowledgment.** The authors thank Dr. Pietro Schino, President of Bari Alzheimer Center, for his grant in the database developing.

## References

1. Plamondon, R., O'Reilly, C., Galbally, J., Almaksour, A., Ancquetil, É.: Recent developments in the study of rapid human movements with the kinematic theory: applications to handwriting and signature synthesis. Pattern Recogn. Lett. (2012, in press). Preprints available on-line at http://dx.doi.org/10.1016/j.patrec.2012.06.004
2. Plamondon, R.: A kinematic theory of rapid human movements: part I: movement representation and generation. Biol. Cybern. **72**(4), 295–307 (1995)
3. Alimi, A.M., Plamondon, R.: Performance analysis of handwritten strokes generation models. In: 3-rd IWFHR, Buffalo, USA, 25–27 May 1993, pp. 272–283 (1993)
4. Plamondon, R.: Handwriting generation: the delta-lognormal theory. In: 4th International Workshop on Frontiers in Handwriting Recognition, pp. 1–10 (1994)

5. O'Reilly, C., Plamondon, R.: Automatic extraction of sigma-lognormal parameters on signatures. In: 11th International Conference on Frontiers in Handwriting Recognition (2008)
6. O'Reilly, C., Plamondon, R.: Development of a sigma-lognormal representation for on-line signatures. Pattern Recogn. (Special Issue on Frontiers in Handwriting Recognition) **42**, 3324–3337 (2009)
7. Plamondon, R.: Strokes against stroke-stroke for strides. In: 3-rd ICFHR, Bari, Italy, 18–20 September 2012. http://www.icfhr2012.uniba.it/ICFHR2012-Invited_I.pdf
8. Plamondon, R., Djioua, M.: A multi-level representation paradigm for handwriting stroke generation. Hum. Mov. Sci. **25**(4–5), 586–607 (2006)
9. Yanikoglu, B., Kholmatov, A.: SUSIG: an online handwritten signature database, associated protocol and benchmark results. Pattern Anal. Appl. **12**, 227–236 (2009). Springer-Verlag, London, 2008
10. Brigham, E.O.: The Fast Fourier Transform. Prentice Hall, Inc., Englewood Cliffs (1974)

# Clusters Identification in Binary Genomic Data: The Alternative Offered by Scan Statistics Approach

Danilo Pellin and Clelia Di Serio[✉]

University Center of Statistics for the Biomedical Sciences,
Vita-Salute San Raffaele University, Milan, Italy
{pellin.danilo,diserio.clelia}@hsr.it
http://www.cussb.unisr.it

**Abstract.** In many different research area, identification of clusters or regions showing an increment in event rate over a given study area is an important and interesting problem. Nowadays literature concerning scan statistics is quite broad and methods can be subdivided based on dimensional complexity of the study area, assumption on distribution generating the data under the null hypothesis and shape-dimension of the scanning window. The aim of this study is to adapt and apply this methodology to the genomics field taking into account for some peculiarities of these data and to compare its performance to existing method based on DBSCAN algorithm.

**Keywords:** Hotspot · Scan statistics · Binary genomic event

## 1   Introduction

Identification of clusters of events over a certain space or of regions showing an higher event rate of occurrence represents an important issue in many research fields, and in particular in biology. Many different approaches have been proposed within both classical and bayesian frameworks, either parametric or non parametric. Most of the algorithms require to specify a priori the number of clusters or their expected dimension and/or length. Usually setting these parameters strongly affects the results if no information are available. This contribution places the problem of cluster identification within a genomic framework. We will compare two different approaches to investigate genomic binary event clustering, the first being the so called Density-Based Spatial Clustering of Applications with Noise (DBSCAN) [8] and the second based on scan statistics methodology proposed by [6].

Both methods have been originally developed in different research areas. DBSCAN has been implemented as a computer science tool, it can search for arbitrary shaped clusters and does not assume any probabilistic models. Kulldorff scan statistics has been mainly used for cluster localization in epidemiological space-time data. Both have been never applied to the analysis of genomic

© Springer International Publishing Switzerland 2014
E. Formenti et al. (Eds.): CIBB 2013, LNBI 8452, pp. 149–158, 2014.
DOI: 10.1007/978-3-319-09042-9_11

data. Although computational demanding the procedure is completely data-driven since no input parameters are needed. Different probabilistic models have been proposed based on the aim of the study, data type and field of application. In this paper we will focus in particular on Bernoulli model that better describes the nature of experimental data. The aim of this study is to extend these methodology to genomic data, accounting for some specific features of the field and comparing identification performances among different settings.

Our experimental framework includes data from gene therapy.

In particular we consider the process of integration in human hematopoietic stem cells CD34+, of viral vector derived from lentivirus Human immune deficiency virus (HIV) and retrovirus Moloney murine leukemia virus (MLV), the most widely used in gene therapy clinical trials [1,2,5,9]. Thanks to the recent developments of Next-Generation Sequencing (NGS) techniques we are able to retrieve the exact genomic coordinates (IS, integration site) for virus integration position in patients' cells. Localization of common integration sites (CIS), that are regions over the genome with high integration density, represents a crucial feature to improve vector safety and to avoid risk of insertional mutagenesis.

## 2    Methods

### 2.1    DBSCAN

In computational biology literature and in particular in viral integration data analysis different methods has been proposed to identify clusters of event or CIS. In [11] for the first time a non-uniform distribution of IS along the genome was assessed and the concept of cluster referred to genomic was introduces. Authors proposed a definition of CIS based simply on proximity between events. In particular they define a portion of the genome as CIS if in 30, 50, 100 kb (kilo-bases) respectively more than 2, 3, 4 IS were observed. Interval length and event counts were completely independent on data sets size and fixed for all experiment.

New techniques, such as high-throughput data, increased the order of magnitude of genomic events available. There are examples in the literature [3] where more than 30 000 events were obtained, about two order of magnitude more than previous experiments. Due to this difference, a new methods based on DBSCAN was proposed. The key idea of this algorithm is that within a neighbourhood of radius *Eps* centered on each observation belonging to a cluster there must be at least other *MinPts* observations. This method classify observations, also termed points, in three categories: (*core points*) and (*border points*) are observation falling within a cluster respectively presenting more (or equal) or less than *MinPts* points in their neighborhood and (*outer points*) located outside any cluster.

In [3] the setting of parameter *MinPts* depends on data sets size, in particular it was fixed to 2 for total count below 10 000 and to 3 for bigger count. *Eps* was set by means of a Monte Carlo procedure consisting in re-sampling 1,000 times the same amount of experimental data from an in-silico generated library generated under the assumption of uniform distribution along the genome. Using the distribution of distances between each event and the *MinPts*-consecutive ones is possible to control the expected number of cluster under the null hypothesis.

## 2.2 Kulldorff Spatial Scan Statistics for Bernoulli Model

The method proposed by [6] addresses clusters identification problem in a very general manner. As aforementioned many different models have been proposed depending on data type, ranging from Poisson to logistic model for categorical data. In this work, we will focus on Bernoulli model, since integration data are binary events. Kulldorff scan statistics vary according to different size of scanning window, to the possibility to deal with arbitrary but known underlying intensity function under the null hypothesis and to the likelihood ratio statistics. The method consists in moving a window of variable size over the genomic area $G$, defining a collection $\mathcal{Z}$ of zones $Z \subset G$. Conditioning on the observed total number of cases, $X$, the spatial scan statistics $S$ is defined as the maximum likelihood ratio over all possible zone $Z \in \mathcal{Z}$:

$$S = \frac{\max\{L(Z)\}}{L_0} = \max_{Z} \left\{ \frac{L(Z)}{L_0} \right\}. \tag{1}$$

$L(Z)$ correspond to the maximum likelihood for zone $Z$, expressing how likely the observed data are, given that events within $Z$ occurs with an higher rate than outside. $L_0$ is the likelihood under the null hypothesis of equal rate over the whole study area.

For the Bernoulli model let define $N$ as the total amount of Bernoulli trials composing to the whole support $G$, $X$ as the total success or event count over $G$, $n_Z$ and $x_Z$ respectively as the count of trials and success observed within the zone $Z$. Conditioning on a given $Z$, the likelihood function for Bernoulli model is:

$$L(Z, p_Z, q_Z) \propto p_Z^{x_Z} (1 - p_Z)^{n_Z - x_Z} \; q_Z^{X - x_Z} (1 - q_Z)^{(N - n_Z) - (X - x_Z)} . \tag{2}$$

To identify $S$, is necessary to maximize the conditional likelihood for all zones $Z \in \mathcal{Z}$ by means of the following function:

$$L(Z) = L(p_Z, q_Z | Z = Z_j) = \left( \frac{x_Z}{n_Z} \right)^{x_Z} \left( 1 - \frac{x_Z}{n_Z} \right)^{n_Z - x_Z}$$
$$\times \left( \frac{X - x_Z}{N - n_Z} \right)^{X - x_Z} \left( 1 - \frac{X - x_Z}{N - n_Z} \right)^{(N - n_Z) - (X - x_Z)} \tag{3}$$

Under the null hypothesis, corresponding to a constant probability of success over $G$, the 2 reduces to:

$$L_0 = \left( \frac{X}{N} \right)^X \left( \frac{N - X}{N} \right)^{N - X} .$$

for all $Z \in \mathcal{Z}$. To replicate the experiment it is sufficient to sample without replacement $X$ locations against $N$ available [4,7,10]. Relying on a considerable high number of simulations, $M = 1\,000$, is possible to retrieve a Monte Carlo estimation of $S_1^*, S_2^*, \ldots S_M^*$. The p-value associated to observed test statistics $S$ will be defined as $P(S) = [1 + \sum_{i=1}^{M} I(S_i^* \geq S)]/(M+1)$ where $I$ is the indicator function.

In [12] two alternative procedures are proposed to identify eventual secondary cluster non overlapping with the most evident. In this work we will use a sequential approach ensuring a good control of I type error and higher power, consisting in removing from $G$ the zone previously detected and redefining a new collection $\mathcal{Z}^* = \mathcal{Z} \setminus \hat{Z}$ of zones to investigate.

### 2.3    Experimental Data

Reference [3] analyze and compare the profile of retrovirus MLV and lentivirus HIV integrations in human hematopoietic stem cells CD34+ in order to study the behavior of the 2 adopted in the same cell type. To reduce possible technical bias the same laboratory protocol and sequencing platform was adopted.

Genomic DNA was extracted 10 to 12 days after transduction and adapted to the GS-FLX Genome Sequencer (Roche/454 Life Sciences) pyrosequencing platform. MLV (n = 244 879) and HIV (n = 163 755) raw sequence reads were processed through an automated bioinformatic pipeline that eliminated short and redundant sequences, and they were mapped on the UCSC hg18 release15 of the human genome (http://genome.ucsc.edu) to obtain 32 631 and 28 382 unique insertion sites, respectively.

As control data set a library of 11 655 601 weighted for possible mappability issue was generated.

## 3    Results

Both MLV and HIV clusters have been clearly defined, identifying integration hot and cold spots. The control library of 11 655 601 in-silico generated random sites were uniformly distributed, except for centromeric, repetitive, or poorly defined regions.

Since both data sets size are bigger than 10 000, DBSCAN parameter *MinPts* was fixed to 3. Using the Monte Carlo resampling procedure the distributions of distance between *MinPts*-consecutive observations was estimated for both data sets. *Eps* were fixed to first percentile of the distributions, in order to control the number of clusters expected under the null hypothesis. Obtained value for *Eps* were 12 587 bp for MLV and 14 460 bp for HIV. Given these parameters setting the following results were achieved by means of DBSCAN.

In HIV data 2 446 clusters were identified over whole genome, containing 50.6 % (14 369) IS of the total count. Cluster length ranges from 100 to 200500 with a mean value of 19220. The 90 % of HIV clusters were composed of 3–10 IS with a maximum of 110 and on average the distance between consecutive integration in cluster was 3 593 bp.

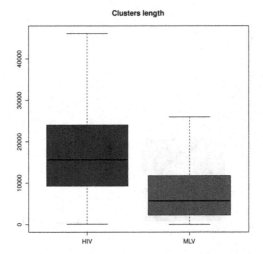

**Fig. 1.** Length distributions of clusters identified by DBSCAN algorithm in HIV and MLV data sets.

For MLV, 3 497 clusters were identify, containing 65.3 % (21 307) ISs of the total count. Clusters length ranges from 19 bp to 78530, with a mean value of 8385. As it can be noticed in Fig. 1 length of clusters identified in MLV data set is substantial smaller than those founded in HIV data set. Regarding the dimensions, as for HIV the main portion of MLV clusters (88 %) are composed of 3–10 IS but the maximum dimension is 42, so considerable smaller than HIV. The average inter-integration distance within MLV clusters is 1 424 bp, suggesting a stronger tendency to aggregate with respect to HIV.

Since no inferential procedure has been proposed for DBSCAN, the cluster were sorted in terms of dimension (IS count) in order to compare with scan statistics results.

Running scan statistics methods with a significance threshold set to $\alpha = 0.05$, 121 and 754 clusters were identified respectively for HIV and MLV.

For HIV derived vector, clusters length vary between a minimum of 3 712 bp to a maximum of 18 206 570 bps, with a mean value of 1 675 751 bps. The count of IS within clusters ranges from 6 to 610, average 92. In general 39 % of IS belong to clusters. The most significant (MSC) over whole genome was located in chromosome 11, interval 64 287 380;66 820 001. The value of statistics $S$ associated is 971.24 and 471 ISs belong to it. The p_value associated to MSC obtained by means of the Monte Carlo procedure is $p - value \leq 0.001$ since the critical value for $S_{0.95}^{*}$ result in 23.612. The most important, in terms of IS, identify using DBSCAN was located in chromosome 11, interval 65 586 752;65 736 062, hence completely included in MSC, see Fig. 3. In addition, it is necessary to highlight that other 22 out of 2446 DBSCAN interesting regions fall completely within MSC margins, including the overall second most relevant. The second most significant cluster, named $MSC_2$ was located on chromosome 17, positions

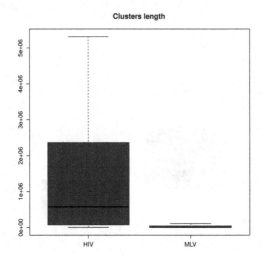

**Fig. 2.** Length distributions of clusters identified by scan statistics algorithm in HIV and MLV data sets.

71 294 851;77 821 445 and 610 IS were included. It corresponds to 38 DBSCAN clusters, including the third most important.

In conclusion, the main part of the difference in total cluster count is attributable to an art fragmentation of scan statistics cluster in more DBSCAN clusters. Despite that, a clear and well defined correspondence in terms of localization and ranking was observed.

For MLV derived vector, 754 clusters were identified using scan statistics approach. As already observed using DBSCAN, clusters for MLV tend to be shorter than HIV (Fig. 2). Length vary from 6 bps to 12 271 839 bps, with a mean value of 222 271 bps, more than seven times smaller then HIV. Also the dimension in term of IS count was significantly different since it varies between 3 and 318, average equal to 20. About 47 % of ISs belong to clusters. The MSC is located on chromosome 20,51 646 845;51 991 770 and contains 89 ISs, see Fig. 4. The corresponding test statistics value $S$ is 198.94, a remarkably significant value compared to the critical value 24.869 calculated using Monte Carlo methods. In MLV data there was no equivalent performance, at least in terms of results ranking, among the two methods. In fact, in genomic portion labeled as MSC, 8 DBSCAN clusters were identified, but the highest in ranking was only 135-th. The second, $MSC_2$, obtained re-running the scan statics procedure after MSC has been remove from the support, was on chromosome 17, interval 26 659 383;26 672 265 and was composed by 39 ISs. It perfectly correspond to the fifth cluster derived from DBSCAN. The most important cluster calculated using DBSCAN is on chromosome 22 27 525 356;27545150, its dimension is 42 and correspond to $MSC_3$ estimated using scan statistics approach.

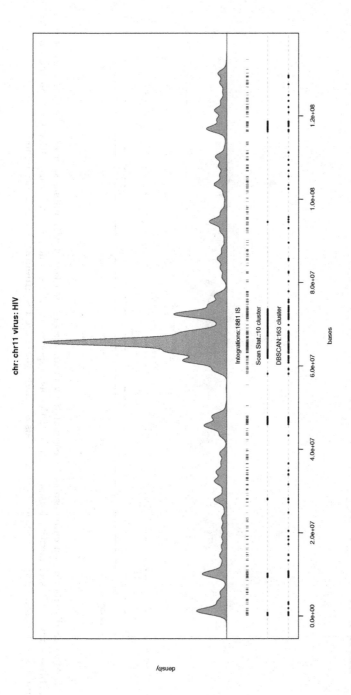

**Fig. 3.** HIV IS distribution on chromosome 11 estimated by means of gaussian kernel with unbiased cross validation bandwidth selection (**blue curve**). Cluster identified using Scan statistics are reported on the second line and using DBASCAN on the third (**black segments**). Most significant cluster for both methods are highlighted (**red segments**) (Color figure online)

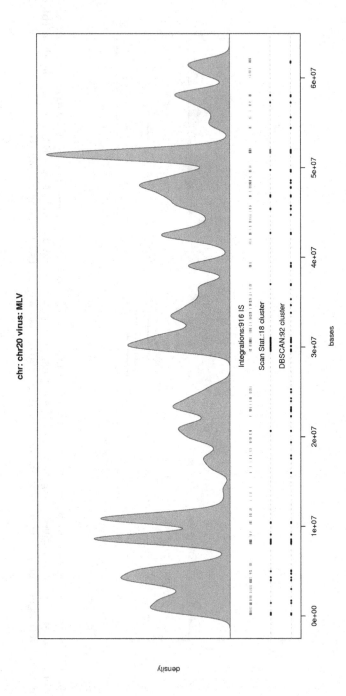

**Fig. 4.** MLV IS distribution on chromosome 20 estimated by means of gaussian kernel with unbiased cross validation bandwidth selection (**blue curve**). Cluster identified using Scan statistics are reported on the second line and using DBASCAN on the third (**black segments**). Most significant cluster for both methods are highlighted (**red segments**) (Color figure online)

# 4  Conclusions

Both methods confirm results previously published revealing deep biological differences in integration process and target sites selection process, when referred to different virus. In particular, speculating only on cluster dimensions and length, our analysis agrees with known preferences of MLV for regulatory element or in general for small genomic interval and HIV for bigger regions corresponding to active coding elements. Despite total cluster count, for HIV data set a substantial overlap between results was observed, in terms of both localization and significance. The intrinsic nature of HIV probably explains this correspondence, since aggregation is weaker in HIV than MLV but affects wider regions, leading to cluster formed by many ISs rewarded by DBSCAN ranking scheme based on dimension. For MLV instead, generally the aggregation tendency is much stronger but related to narrower genomic intervals.

Scan statistics, relying on a probabilistic approach based on success probability estimates and comparison is able to capture this aspect and using hypothesis testing procedure is possible to sort the regions in terms of effect strength instead of dimension. Finally, scan statistics seems to be more flexible than DBSCAN in capturing different clustering behavior and no input parameters, except for significance threshold, is needed. The cost to pay for this unsupervised search is computational time.

# References

1. Aiuti, A., Cattaneo, F., Galimberti, S., Benninghoff, U., Cassani, B.: Gene therapy for immunodeficiency due to adenosine deaminase deficiency. N. Engl. J. Med. **360**, 447–458 (2009)
2. Cartier, N., Hacein-Bey-Abina, S., Bartholomae, C., Veres, G., Schmidt, M., et al.: Hematopoietic stem cell gene therapy with a lentiviral vector in X-linked adrenoleukodystrophy. Science **326**, 818–823 (2009)
3. Cattoglio, C., Pellin, D., Rizzi, E., Maruggi, G., Corti, G., Miselli, F., Sartori, D., Guffanti, A., Di Serio, C., Ambrosi, A., De Bellis, G., Mavilio, F.: High-definition mapping of retroviral integration sites identifies active regulatory elements in human multipotent hematopoietic progenitors. Blood **116**, 5507–5517 (2010)
4. Dwass, M.: Modified randomization tests for nonparametric hypothesis. Ann. Math. Stat. **28**, 181–187 (1957)
5. Hacein-Bey-Abina, S., Le Deist, F., Carlier, F., Bouneaud, C., Hue, C., et al.: Sustained correction of X-linked severe combined immunodeficiency by ex vivo gene therapy. Engl. J. Med. **346**, 1185–1193 (2002)
6. Kulldorff, M.: A spatial scan statistic. Commun. Stat. Theory Methods **26**, 1481–1496 (1997)
7. Loader, C.R.: Large-deviation approximation to the distribution of scan statistics. Ann. Appl. Probab. **23**, 751771 (1991)
8. Martin, E., Kriegel, H.P., Sander, J., Xu, X.: A density-based algorithm for discovering clusters in large spatial databases with noise. In: Proceedings of 2nd International Conference on Knowledge Discovery and Data Mining (KDD-96), Institute for Computer Science. University of Munich (1996)

9. Mavilio, F., Pellegrini, G., Ferrari, S., Di Nunzio, F., Di Iorio, E., et al.: Correction of junctional epidermolysis bullosa by transplantation of genetically modified epidermal stem cells. Nat. Med. **12**, 1397–1402 (2006)

10. Turnbull, B., Iwano, E.J., Burnett, W.S., Howe, H.L., Clark, L.: Monitoring for cluster of disease: application to leukemia incidence in upstate New York. Am. J. Epidemiol. **132**, S136–S143 (1990)

11. Wu, X., Luke, B.T., Burgess, S.M.: Redefining the common insertion site. Virology **344**, 292–295 (2006)

12. Zhang, Z., Assuno, R., Kulldorff, M.: Spatial scan statistics adjusted for multiple clusters. J. Probab. Stat. **11** (2010)

# Special Session:
# Knowledge Based Medicine

# Reverse Engineering Methodology for Bioinformatics Based on Genetic Programming, Differential Expression Analysis and Other Statistical Methods

Corneliu T.C. Arsene[✉], Denisa Ardevan, and Paul Bulzu

Solutions of Artificial Intelligence Applications,
45 Vlahuta Street, Cluj-Napoca, Romania
ArseneCorneliu@yahoo.co.uk,
{Corneliu.Arsene,Denisa.Ardevan,
Paul.Bulzu}@saia-institute.org

**Abstract.** This paper presents a robust automatic modelling of microRNA (miRNA) dynamics which has at core a Genetic Programming (GP) Reversing Ordinary Differential Equations Systems (RODES) methodology which was developed before and which is enhanced herein and consists of four steps: (1) smooth and fit the miRNA experimental data which is enhanced in this paper with other types of statistical analyses such as gene differential expression analyses, (2) decomposition of the transcription network or ODEs system, (3) automatically discovering the structure of networks or their ODEs system model by using GP and automatically estimating parameters of the ODEs systems models, (4) identification of the biochemical and pharmacological mechanisms. All four steps are paramount to the GP RODES methodology, which has been already applied in bioinformatics, while in this paper it is underlined a robust set of procedures for implementing the first step of the four step GP RODES framework together with an application of the GP RODES on a real miRNA dataset. Specifically it is highlighted a robust method to noise for fitting omics experimental input data, which consists of the Smoothing Spline Regression (SSR) algorithm based on the Generalized Cross Validation (GCV) criterion. The differential expression analysis of a dataset of 837 mirRNAs genes (GEO accession: GSE35074) is achieved by using various statistical methods. Furthermore, a GP algorithm (i.e. GP TIPS software) is used on the SSR and GCV fitted miRNA200a (microRNA200a), miRNA200b and miR-NA424, which were selected from a larger group of identified differential expressed miRNA genes and with the scope of predicting accurately the miRNAs derivatives with regard to time. While the computational findings are in agreement with the ones from the initial study by Moes et al. (2012) [21], this paper develops an enhanced GP RODES methodology which can be further applied to bioinformatics.

**Keywords:** GP RODES methodology · Genetic programming · Differential expression analysis · Smoothing spline regression · MicroRNA

© Springer International Publishing Switzerland 2014
E. Formenti et al. (Eds.): CIBB 2013, LNBI 8452, pp. 161–177, 2014.
DOI: 10.1007/978-3-319-09042-9_12

# 1    Introduction

The purpose of any biochemical networks reverse engineering method [1–5] is to determine the mechanisms, structure and interactions of the respective networks by inferring their properties from data and knowledge rather than enforcing a mathematical model. By using reverse engineering, the need for having initially a mathematical model which describes the structure or dynamics of the networks can be avoided [6].

In the last decade, many reverse engineering studies [7] attempted to determine the structure and interactions of these networks by using boolean networks [8], probabilistic models [9], evolutionary computation [10] or non-linear Ordinary Differential Equations (ODEs) [11]. Genetic Programming (GP) inferring from genomics input data [1, 6, 12] has recently shown very good perspectives for determining the characteristics of such biochemical networks, as for example gene regulatory networks [13].

A robust methodology is the GP Reversing Ordinary Differentially Equation System (GP RODES) [6] which can automatically determine ODE systems from genomics data by using the GP. In comparison to other methods based on evolutionary computation, GP RODES can deal with incomplete data (i.e. either missing variables or variables with missing values) in the high-throughput time-series omics datasets. It can also be coupled with Neural Network (NN) RODES [6] when some partial knowledge is known about the gene regulatory network and the input omics dataset. GP RODES is comprised of four successive procedures, each of them addressing a different problem in an efficient and effective manner. Various GP algorithms or software applications can be used to implement the GP part of GP RODES: Tree GP (e.g. GPLAB software, GPTIPS software, ECJ software), Linear GP (e.g. Discipulus software), Gene Expression Programming (e.g. GeneXproTools software).

This paper is structured as follows: first the GP RODES methodology is presented, then numerical results are shown on a real genomics dataset and the paper ends with conclusions.

# 2    Enhanced GP RODES Methodology

The GP RODES methodology starts with the omics time-series data. It can actually be applied to all types of high throughput time-series data for which an ODE can be considered as an adequate model. In this case, the main focus lies on the application of the methodology on the microRNA (miRNA) omics data.

The result of GP RODES is an ODE system, $dX/dt = g(X)$ with n equations, one for each variable or network node $X_i$, $i = 1,...,$ n and t is time. The GP RODES methodology applied on a generic miRNA dataset for obtaining reverse engineered transcription networks includes the following steps:

1.    smooth and fit the miRNA experimental data that can be enhanced with various statistical methods for differential expression analysis of omics data, clustering methods, genomic correlation measurements, etc.;

2.  decomposition of the transcription network or ODEs system [14];
3.  automatically discovering the structure of the networks or their ODEs system model and automatically estimating the parameters of the ODEs by using GP;
4.  identification of the biochemical and pharmacological mechanisms involved;

The decomposition of transcription networks or ODEs system models is described in [14] in the context of GP RODES together with the identification of biochemical and pharmacological mechanisms. Here, the main focus will be on the first step of the GP RODES methodology. This first step includes preparation of the input data together with various types of statistical analyses which can be applied on this omics input data, such as the analysis of differentially expressed genes or gene clustering. Before the core of GP RODES comes in place (i.e. steps 2, 3 and 4), it is mandatory to have reliable genomics (miRNA) input data as well as to gain knowledge regarding the omics dataset by using various types of analyses for the purpose of ODEs determination, which describe the interactions within a given omics (miRNA) dataset.

The general ODE model of miRNA transcription is shown in the Eq. (1), which is used to calculate the rate of change of a miRNA:

$$\frac{dmiRNA}{dt} = \beta_0 + \beta \cdot P - \gamma \cdot miRNA. \tag{1}$$

where $\beta_0$ is the basal miRNA transcription rate (which could be zero), $\beta$ represents the maximal transcription rate, $\gamma$ is the miRNA degradation constant, $P$ is a probability function describing miRNA production regulation by other nodes of the drug gene regulatory network, Transcription Factors and drug related compounds.

The probability function $P$ is increasing for activation and decreasing for inhibition. $P$ could be calculated if the concentration profiles for all regulators and other parameters involved are known. However this is not the case in most real-world situations. Therefore in [14, 15] a method was developed to estimate $P$ when information is incomplete. The respective work in [14, 15] introduced the notion of regulome which defines the set of regulatory factors (e.g. TFs) which can influence the miRNA transcription. Furthermore, in the case of missing information in the respective regulome, the regulome probabilities (i.e. depending on TFs) were introduced which formed regulome functions. These functions could be estimated by means of GP RODES by using certain types of functions which can accurately describe regulatory interactions in empirical data [14, 15].

The GP part of GP RODES, which is used to determine the ODEs, can be described as follows:

1.  Initialize a population of randomly generated individuals coding mathematical models relating inputs $X_i$ to outputs $dX_i/dt$.
2.  Repeat the following steps until an acceptable solution is found or some other stopping condition is met (e.g. a maximum number of generations is reached):

    (a) Run each individual and ascertain its fitness. The fitness is computed using the mean squared error and it depicts how well an individual maps the input data $X_i$ to the output data $dX_i/dt$.

(b) Randomly select several individuals from the population (usually, between 1 and 4 individuals) with a fitness based probability in order to participate in genetic operations.

(c) Create new individuals by applying genetic operations. As an example, from 4 individuals selected at the previous step, take two individuals as winners and the other two as losers. Copy the two winner individuals and transform them probabilistically by applying several genetic operators: i. crossover and/ or; ii. mutation; iii. reproduction.

3.  Return the best-so-far individual.

The enhanced GP RODES methodology can be summarized in Fig. 1. When GP RODES has no missing variables then it can use as input variables values $X_i(t)$, i = 1,..., n at discrete time points t = 1,..., T. If variables are missing, GP RODES can also use as inputs the regulomes probabilities $P_j(t)$, j = 1,..., m as in [14, 15]. GP RODES predicts the first order derivative of the miRNAs of interest. This way GP RODES performs reverse engineering on individual algebraic decomposed ODEs instead of reverse engineering on the whole system of ODEs.

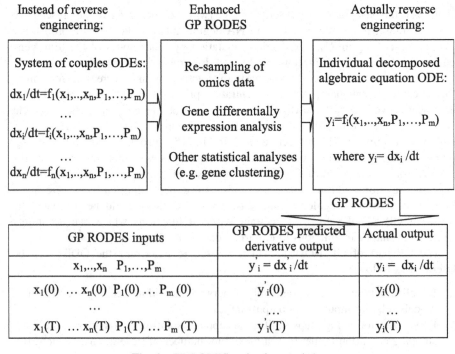

**Fig. 1.** GP RODES main characteristics.

## 2.1 Data Fitting with Smoothing Spline and Time Derivative Computing

Machine learning algorithms (e.g. GP, NN) [16] require enough input points in order to be able to use the respective learning algorithms. In order to apply the GP RODES methodology on the experimental miRNA data, a sufficient number of experimental points must be available for a miRNA. Frequently this is not the case and hence it is necessary to use interpolation methods which fit a function of the available experimental points. By sampling this function, the necessary number of points can be obtained. The smoothing splines methodology is one of the interpolation methods that are robust to noise in the experimental data. Therefore, the method of choice here is the Smoothing Spline Regression (SSR) with Generalized Cross-Validation (GCV) criterion [17]. The determination of smoothing parameter $\lambda$, which controls the fitted function $f$, is of interest. The smoothing parameter can make the function $f$ smoother or closer to the data.

$$\frac{1}{n}\sum_{i=1}^{n}\left(y_i - f(x_i)\right)^2 + \lambda \int_0^1 \left(f^{(l)}\right)^2 dx. \tag{2}$$

where $n$ is the total number of experimental points of a gene, $i$ is an index describing the number of experimental points, $x_i$ is the experimental data, $\int_0^1 \left(f^{(l)}\right)^2 dx$ is a penalization term that regulates the amount by which $f$ is permitted to vary relative to a polynomial model, $l$ is the order of the derivative.

Equation (2) defines the Penalized Least Squares (PLS) which is minimized so that:

$$\int_0^1 \left(f^{(l)}\right)^2 < \beta. \tag{3}$$

where $\beta$ is a constant.

The GCV estimate of $\lambda$ minimizes the function:

$$GCV(\lambda) = \frac{\frac{1}{n}\left\|(I - D(\lambda))y\right\|^2}{\left(\frac{1}{n}tr(I - D(\lambda))\right)^2}. \tag{4}$$

where $tr$ is the trace of a matrix, $I$ is the identity matrix of size n, $D(\lambda)$ is the influence matrix with the property that multiplied by $y$ will give the fitted points [17].

The fitted data produces a function that can be used to resample the miRNA time series data. Once the resampling is done, the time derivative of the resampled time data must be calculated since this derivative will be used as input for GP RODES.

Careful consideration has to be taken as the derivation of a signal affected by low noise can increase the level of noise in the signal derivative. Therefore, derivation is used here for exploratory purposes first, based on a centered difference formula with two points (Eq. 5) and subsequently, by using an eleven-point centered difference approximation [18] (Eq. 6).

$$f'(x_0) = \frac{f(x_0 + h) - f(x_0 - h)}{2h}.$$

(5)

$$f'(x_0) = \frac{\begin{array}{c} 3f(x_0 - 4h) - 32f(x_0 - 3h) + 168f(x_0 - 2h) - 672f(x_0 - h) \\ \hline +672f(x_0 + h) - 168f(x_0 + 2h) + 32f(x_0 + 3h) - 3f(x_0 + 4h) \end{array}}{840h}.$$

(6)

where $x_0$ is the point where the first derivative is calculated, h is a certain step.

## 2.2  Temporal Differentially Expressed Genes

Statistical analysis methods can be applied on miRNA data with the aim of discovering differentially expressed genes. These statistical methods can be incorporated in the first step of the 4-step GP RODES producing an enhanced framework.

A commonly used method involves the application of a certain statistical test for each gene (e.g. t-test) thus obtaining a probability $p_g$ that under the null hypothesis (i.e. not statistical significant) follows the expression $\Pr(p_g < \lambda) = \lambda$. A low $p_g$ value can be seen as evidence that the null hypothesis may not be true, and therefore lead to the conclusion that the gene is differentially expressed. The t-test is defined below where it is compared for example the gene mean against a mean value of 0:

$$t = \frac{\bar{x}}{s/\sqrt{n}}.$$

(7)

where $t$ is the t-value, $\bar{x}$ is the mean calculated over all the expression points for a gene, $n$ is the number of time points for a gene, $s$ is the variance of the respective gene.

Considering that the gene has 97 expression points corresponding to 97 consecutive time points, this would count to 96 degrees of freedom in the t-test. Corroborated with a two-tailed t-test and a significance level of 0.01 (i.e. p-value), it would correspond to a critical t-value of 2.63. Therefore t-values above 2.63 would correspond to differentially expressed genes while t-values under 2.63 would not be considered as indicators of differentially expressed genes.

However, another type of t-test would be to make a comparison of Eq. (7) with a high numerical value (e.g. 100000) for each gene. In the case of this type of t-test, small t-test values would correspond to differentially expressed genes and big t-test values would correspond to non-differentially expressed genes.

Time course studies of differentially expressed genes can provide information regarding the differentiation of a set of genes by taking into account the entire shape of the gene expression function of time. The time course method, based on t and F statistics and implemented by the software Edge [19], realizes time course studies of genes. This method initially fits a mathematical model to the gene expression under the null hypothesis by imposing constraints on the shape of the fitted curve, and subsequently fits a different mathematical model under an alternative hypothesis by using a general parametrical form based on cubic splines for the fitted curve. Then the sum of squares of the differences between the fitted values and the given values for the null hypothesis model $S_i^0$ and for the alternative hypothesis model $S_i^1$ form the statistics for gene $i$ [19]:

$$F_i = \frac{S_i^0 - S_i^1}{S_i^1}. \tag{8}$$

Based on the $F_i$ value for each gene, the statistical measures p-value and the Q-value can be calculated for each gene by determining the probability that the null hypothesis model for a gene is more statistically significant than the alternative hypothesis model.

The Bayesian Estimation of Temporal Regulation (BETR) algorithm is a time course study [20] which works by taking into consideration the correlation between the different time points of a gene and calculating the respective correlation terms. Based on these, the probability of a differentially expressed gene is determined by using a treatment gene group and a control gene group.

$$P(g = 1 \mid D_{tn}) = \frac{P(D_{tn} \mid g = 1)\, P(g = 1)}{(1 - P(g = 1))\, P(D_{tn} \mid g = 0) + P(g = 1)\, P(D_{tn} \mid g = 1)}. \tag{9}$$

where $g$ equals 1 denotes a differentially expressed gene, $D_{tn}$ is the difference in gene expression between the treatment and the control gene groups, $P(g = 1 \mid D_{tn})$ is the probability of a differentially expressed gene given the difference in gene expressions between the two control groups, $P(D_{tn} \mid g = 1)$ is the probability of having a difference between the two gene expression control groups given a differentially expressed gene, $P(D_{tn} \mid g = 0)$ is the probability of having a difference between the two gene expression control groups given a not differentially expressed gene, $P(g = 1)$ is the probability of differentially expressed genes.

The calculation of the differentially expressed genes $g$ is realized through a fast iterative process which starts with an initial estimation of $P(g = 1)$ based on which the probability $P(D_{tn} \mid g = 1)$ is determined [20] and then a first estimation of $g$ is obtained. This process is repeated until $g$ becomes stable and the differentially expressed genes are identified. The differentially expressed genes are the ones which have the probability $P(g = 1 \mid D_{tn})$ higher than an imposed value.

## 3 Results

The methods presented in the previous theoretical sections and which are now included in the first step of the GP RODES framework are applied to a time-resolved microarray data (GEO accession: GSE35074) of Epithelial to Mesenchymal Transition (EMT). This study primarily focuses on the potential effects of SNAI1 transcription factor on a set of 837 miRNAs extracted from a culture of transformed MCF7 breast carcinoma cells [21]. The experimental data points of the 837 miRNAs are given at time points 0h, 4h, 8h, 12h, 24h, 72h and 96h. For each miRNA there are four technical and three biological replicates. Between 72h and 96h is achieved the Mesenchymal stable state for the differentially expressed miRNAs because of the influence of the SNAI1 transcription factor.

## 3.1    Data Fitting and Time Derivative Computing

In Fig. 2 the SSR with GCV fitted curve is shown for miRNA200a, miRNA200b and miRNA424 considering the mean of their four technical and three biological replicates. SSR with GCV is applied to the entire set of miRNAs. Since the derivative of the miRNA function of time (dmiRNA/dt) is required by GP RODES, it is done for comparison purposes with two methods: an eleven-point finite difference approximation and a derivative based on two points centered difference formula. Both methods are applied to the entire miRNAs dataset. In Fig. 3 the derivative of miRNA200a, miRNA200b and miRNA424 are shown, calculated using the two formulas.

Over the entire miRNA dataset, the maximum differences between the two methods of derivative calculation with respect to time are shown in Fig. 4. Some differences can be noticed between the two methods particularly in the first 10 time points out of the 97 points. For example, in Fig. 3, a bigger difference can be observed for the first two time points in the derivative of miRNA200a of 0.1 and a smaller difference for the calculated derivative of miRNA200b of 0.006.

As especially the concern is on the values of the derivative towards the middle and the end of the time of study, the two point finite difference approximation which is similar with the 11-eleven point method for the respective time period is chosen to be used in the calculation of the derivative of the miRNAs dataset and subsequently in GP RODES.

## 3.2    Results of Time Course Study

It is important to find the actual differentially expressed genes from a larger set which involves the identification of directly and indirectly affected miRNAs relative to one or more transcription factors. In this study, four algorithms are used: a time course study based on BETR method [20], the methods implemented by Edge software which could be either a static or a time course study, as well as a specific t-test explained above.

For exploratory purposes, the methods for differentially expressed genes analysis can be applied on the experimental miRNA data, the fitted miRNA data or the fitted miRNA data at the experimental time points.

The differential expression analysis is realized first with BETR implemented with the MultiExperiment Viewer (MEV) [22] software. In the analysis made with BETR, there were used three experimental biological replicates/groups. The numerical results in Fig. 5 showed a top-list of 27 miRNAs being differentially expressed, while the extended list obtained with BETR had many similarities with a list of 350 experimentally validated miRNAs that are influenced by SNAI1 transcription factor and which 350 experimental validated gene list was obtained using the MetaCore application's database [23]. Specifically, the miRNA200 group was identified as being differentially expressed within the reduced 27 miRNAs list together with other experimentally confirmed miRNAs (e.g. miRNA424, miRNA221, etc.).

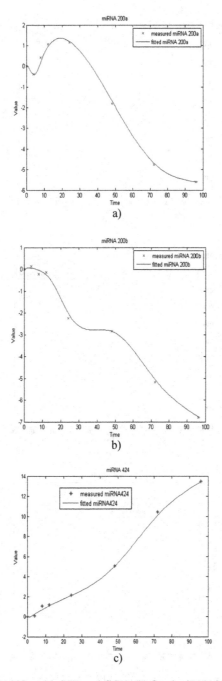

**Fig. 2.** (a) Fitted miRNA200a with SSR and GCV; (b) fitted miRNA200b with SSR and GCV; (c) fitted miRNA424 with SSR and GCV.

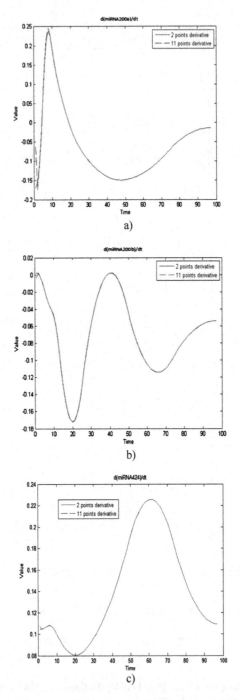

**Fig. 3.** (a) Derivative of fitted d(miRNA200a)/dt; (b) derivative of fitted d(miRNA200b)/dt; (c) derivative of fitted d(miRNA424)/dt.

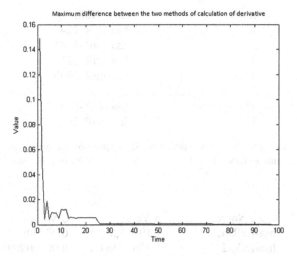

**Fig. 4.** Maximum difference between the derivatives calculated with the two methods (two points and eleven points difference approximation) over the entire miRNAs dataset.

| Rank | Gene Name |
|------|-----------|
| 1 | hsa-miR-221 |
| 2 | hsa-miR-200b |
| 3 | hsa-miR-424* |
| ......................................... | |
| 7 | hsa-miR-342-3p |
| 8 | hsa-miR-200a |
| ..................................... | |
| 27 | hsa-miR-200c |

**Fig. 5.** 27 miRNAs identified with BETR using the three experimental biological replicates.

These findings are also in agreement with the ones reported in the initial study by Moes et al. (2012) in [21] which addressed firstly this miRNA dataset (GEO accession: GSE35074).

A second differentially analysis was carried out with software Edge and the static analysis implemented by the software. Two groups were formed, Group A consisting of the fitted data at times 0h, 1h, 2h, 3h, 4h, 5h and Group B consisting of the fitted data at times 94h, 95h, 96h. A static analysis was implemented with Edge and the ranked list of differentially expressed miRNAs showed again similarities with the previous result (Fig. 6).

A third analysis implemented herein was a time course study realized with Edge on the experimental data for the mean of the four technical replicates for each biological replicate. There were again three biological replicates used in the time course study for each miRNA.

The numerical results (Fig. 7) clearly identified a set of 34 miRNAs as being differentially expressed in conformity with the previous results reported here and in

| Rank | Gene Name |
|------|-----------|
| 1 | hsa-miR-424 |
| 2 | hsa-miR-424* |
| 3 | hsa-miR-221 |
| 4 | hsa-miR-200b |
| ......... | ......... |
| 11 | hsa-miR-503 |
| 12 | hsa-miR-200a |

**Fig. 6.** miRNAs identified in a static differentially expressed analysis with Edge software considering fitted points at times 0 h, 1 h, 2 h, 3 h, 4 h, 5 h (Group A) and 94 h, 95 h, 96 h (Group B).

| Rank | Gene Name | p-Value | Q-value |
|------|-----------|---------|---------|
| 1 | hsa-miR-424 | 1.196172e-05 | 0.000909091 |
| 2 | hsa-miR-221 | 1.196172e-05 | 0.000909091 |
| 3 | hsa-miR-200b | 1.196172e-05 | 0.000909091 |
| 4 | hsa-miR-183 | 1.196172e-05 | 0.000909091 |
| 5 | hsa-miR-503 | 1.196172e-05 | 0.000909091 |
| 6 | hsa-miR-200c | 1.196172e-05 | 0.000909091 |
| 7 | hsa-miR-424* | 1.196172e-05 | 0.000909091 |
| 8 | hsa-miR-200a | 1.196172e-05 | 0.000909091 |
| ...... | ...... | ...... | ...... |
| 34 | hsa-miR-151-3p | 0.003923445 | 0.096470588 |

**Fig. 7.** 34 miRNAs identified in a time course study with Edge software using three biological replicates.

the literature concerning experimental studies. The ranking was made based on the p-value statistical measure while the Q-value is also shown.

In conclusion the numerical results of the differentially expression analyses showed a consistent degree of similarity among the time course studies and the static analysis. These differentially expression analyses can be combined with other types of statistical analyses such as clustering (e.g. Self Organizing Maps, etc.) for the purpose of identification of the traits of the omics datasets.

The final differentially expression analysis is realized for the mean of the biological and the technical replicates from above and with t-test implemented with the same software MEV. By using the fitted data with SSR and GCV, a t-value for each gene is calculated by comparing the mean of a gene with a high numerical value (e.g. 100000) on the grounds that genes that are differentially expressed may have the mean value bigger than the means of the other genes while the standard deviations being smaller for the differentially expressed genes: this would result in small t-test values for the differentially expressed genes. Once again in this experiment we are interested in small t-test values as a way of detecting the differentially expressed genes by comparison with a high numerical value. The numerical results in Fig. 8 show that most of the genes which clustered at the bottom of the ordered list of genes with

| Rank | Gene Name |
|------|-----------|
| 1 | hsa-miR-424 |
| 2 | hsa-miR-429 |
| 3 | hsa-miR-200a |
| 4 | hsa-miR-424* |
| 5 | hsa-miR-221 |
| 6 | hsa-miR-181b |
| 7 | hsa-miR-200b |
| 8 | hsa-miR-451 |
| ................................ | |
| 40 | hsa-miR-380 |

**Fig. 8.** 40 miRNAs identified with t-test using the mean of the three biological and technical replicates and a comparison with a high numerical value (e.g. 100000).

respect to the t-values had many similarities with the lists obtained in the other experiments and includes the group miRNA200a.

This analysis gives appropriate results also for the genes which are affected by SNAI1 in the EMT and are both up-regulated and down-regulated and would potentially have the mean value very small (i.e. closed to 0). In this last case the standard deviation of the respective genes are represented by high values and the comparison with a high numerical value would still result in small t-test values as above and would successfully determine the differentially expressed genes.

## 3.3   Structure and Parameters Discovery from the GSE35074 miRNA Time Series Data with GP RODES

GP RODES is used to obtain an ODE for dmiRNA200a/dt. Specifically from the suite of GP software available, the genetic programming software GPTIPS (Genetic Programming & Symbolic Regression Toolbox for MATLAB) [23] is used with the following parameters: (1) Population size 500, (2) Number of generations 2000, (3) Number of trees in tournament 7, (4) Maximum tree depth 12, (5) Maximum number of genes 2, (6) Probability of tree mutation 0.1, (7) Probability of tree crossover 0.85, (8) Probability of tree direct copy 0.05, (9) Function set $\{+, -, *, /\}$. The elitism selection was present also as an initial setting in the software, each individual could have a maximum number of 2 genes and no duplicates were allowed at the initialization stage and the operator "/" was protected against division by zero.

The ODE, which might be expected to obtain for miRNA200a with GP RODES and with a basal miRNA transcription rate $\beta_0$ of zero, has the form:

$$\frac{d(miRNA200a)}{dt} = ks * P(SNAI1, \ldots) - kd * miRNA200a \tag{10}$$

where $k_s$ is the miRNA200a production constant, $k_d$ is the miRNA200a degradation constant, *SNAI1* is one of the transcription factors which regulates miRNA200a.

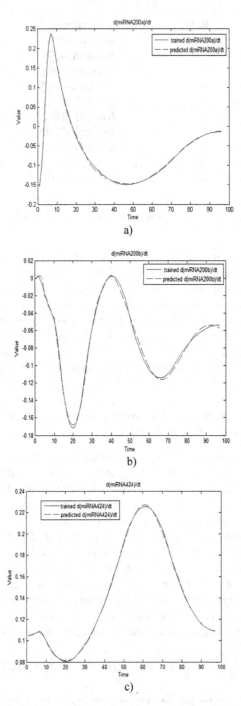

**Fig. 9.** (a) Measured and predicted d(miRNA200a)/dt; (b) measured and predicted d(miR-NA200b)/dt; (c) measured and predicted d(miRNA424)/dt.

In Eq. (10) the probability function $P$ describing miRNA200a production regulation by other TFs and drug related compounds depends in special on the transcription factor SNAI1 based on the experimental information. The miRNA200a was found by the differentially expression analyses to be influenced directly in EMT by SNAI1 confirming also the experimental results [24]. The complexity of the equation which describes d(miRNA200a)/dt can be controlled by changing for example the tree depth in the GPTIPS software so that to obtain a simpler or a more complex equation for a given level of accuracy in the numerical results.

In Fig. 9(a) is shown the measured and the predicted d(miRNA200a)/dt obtained with GPTIPS where the Root Mean Square Error (RMSE) is 0.0033749. In Fig. 9(b) is shown the measured and the predicted d(miRNA200b)/dt where the RMSE is 0.0032146. In Fig. 9(c) is shown the measured and the predicted d(miRNA424)/dt where the RMSE is 0.00091695.

However, in the examples presented herein, more investigation in the structure of the ODEs might be required as described in [15].

By integrating the ODE determined with respect to time on the interval [0h 96h] for example for d(miRNA424)/dt, it is obtained the predicted miRNA424 values for the respective time interval. Finally, once the discovering of the structure of the networks or their ODEs system model has been done, it can take place the identification and the study of the biochemical and pharmacological mechanisms involved, which is the last step of the 4-step GP RODES methodology.

## 4  Conclusions

The enhanced GP RODES methodology was applied successfully to a high-throughput time-series omics dataset of miRNAs which are influenced in the EMT by transcription factors, especially SNAI1.

First, there was an interest to incorporate statistical analyses in the first step of the four-step GP RODES framework, which could provide useful information or knowledge regarding the interactions present in a miRNA dataset. A series of differential expression analyses were carried out on a miRNA dataset based on BETR algorithm, two methods implemented by Edge software and a specific t-test calculation. The lists of differentially expressed genes, which were found with the statistical tests, confirmed the expectations based on the experimentally validated results as well as the ones reported in the initial study by Moes et al. (2012) in [21] which addressed initially this miRNA dataset (GEO accession: GSE35074). It can be concluded that various combinations of statistical tests can be incorporated at the first step of the four-step GP RODES with the aim of alleviating the work of biomedical researchers with respect to deriving the needed information regarding the differentially expressed genes.

Finally, by using a GP algorithm (i.e. GP TIPS software) applied on data obtained with a robust fitting method SSR with GCV criterion, the aim was to predict accurately and to determine reliable ODEs, which describe how miRNA200a, miRNA200b and miRNA424 vary in time function of the inhibitory and stimulatory biological molecules. This information can be useful in the biomedical research field in determining the drugs which can tackle serious diseases such as cancer disease. Further work will involve the comparison of the GP RODES with other methodologies or computational models based for example on NNs [6, 25–27].

**Acknowledgments.** This work was supported by a grant of the Romanian National Authority for Scientific Research, CNCS – UEFISCDI, project number PN-II-PT-PCCA-2011-3.1-1221, named IntelUro. The author would also like to thank for their support to Dr. Alexandru Floares and Dr. Adriana Birlutiu.

# References

1. Crombach, A., Wotton, K.R., Cicin-Sain, D., Ashyraliyev, M., Jaeger, J.: Efficient reverse-engineering of a developmental gene regulatory network. PLoS Comput. Biol. **8**(7) (2012)
2. D'haeseleer, P., Liang, S., Somogyi, R.: Genetic network inference: from co-expression clustering to reverse engineering. Bioinformatics **16**, 707–726 (2000)
3. Jaeger, J., Crombach, A.: Life's attractors: understanding developmental systems through reverse engineering and in silico evolution. Adv. Exp. Med. Biol. **751**, 93–119 (2012)
4. Garnder, T.S., Faith, J.J.: Reverse-engineering transcription control networks. Phys. Life Rev. **2**, 65–88 (2005)
5. Rockman, M.V.: Reverse engineering the genotype-phenotype map with natural genetic variation. Nature **456**(7223), 738–744 (2008)
6. Floares, A.G.: Toward personalized therapy using artificial intelligence tools to understand and control drug gene networks. In: Ramov, B. (ed.) New trends in technologies. Intech, Rijeka (2010). ISBN 978-953-7619-62-6
7. Hecker, M., Lambeck, S., Toepfer, S., van Someren, E., Guthke, R.: Gene regulatory network inference: data integration in dynamic models-a review. BioSystems **96**, 86–103 (2009)
8. Bornholdt, S.: Boolean network models of cellular regulation: prospects and limitations. J. R. Soc. Interface **5**, S85–S94 (2008)
9. Nariai, N., Kim, S., Imoto, S., Miyano, S.: Using protein–protein interactions for refining gene networks estimated from microarray data by Bayesian networks. In: Proceeding of the Pacific Symposium on Biocomputing, pp. 336–347 (2004)
10. Vilela, M., Chou, I.C., Vinga, S., Vasconcelos, A.T., Voit, E.O., Almeida, J.S.: Parameter optimization in S-system models. BMC Syst. Biol. **16**(2), 35 (2008)
11. Voit, E.O.: Modelling metabolic networks using power-laws and S-systems. Essays Biochem. **45**, 29–40 (2008)
12. Spieth, C., Hassis, N., Streichert, F.: Comparing mathematical models on the problem of network inference. In: Proceeding of the 8th Annual Conference on Genetic and evolutionary computation (GECCO 2006), Washington, USA, pp. 279–285 (2006)
13. Bar-Joseph, Z., Gitter, A., Simon, I.: Studying and modeling dynamic biological processes using time-series gene expression data. Nature **13**, 552–564 (2012)
14. Floares, A., Birlutiu, A.: Reverse engineering networks as ordinary differential equations systems. In: Floares, A. (ed.) Computational Intelligence. NOVA Science Publishers, New York (2012)
15. Floares, A.G., Luludachi, I.: Automatically inferring the dynomics and regulomics of transcription networks with unknown transcription factors and microRNAs regulators using GP RODES. In: Kasabov, N. (ed.) Springer Handbook of Bio-/Neuroinformatics, pp. 311–326. Springer, Heidelberg (2014)

16. Floares, A.G.: Computational intelligence tools for modeling and controlling pharmacogenomic systems: genetic programming and neural networks. In: Proceedings of the 2006 IEEE World Congress on Computational Intelligence, Vancouver, CA. IEEE Press (2006)
17. Wang, Y.: Smoothing Splines: Methods and Application. CRC Press, Boca Raton (2011)
18. Stewart, S.: Calculus, 7th edn. Brooks/Cole, Pacific Grove (2012)
19. Storey, J.D., Xiao, W., Leek, J.T., Tompkins, R.G., Davis, R.W.: Significance analysis of time course microarray experiments. PNAS 102(36), 12837–12842 (2005)
20. Aryee, M.J., Gutierrez-Pabello, J.A., Kramnik, I., Maiti, T., Quackenbush, J.: An improved empirical Bayes approach to estimating differential gene expression in microarray time-course data: BETR (Bayesian Estimation of Temporal Regulation). BMC Bioinform. 10, 409 (2009)
21. Moes, M., Le Bechec, A., Crespo, I., Laurini, C., Halavatyi, A., Vetter, G., Del Sol, A., Friedercih, E.: A novel network integrating a miRNA-203/SNAI1 feedback loop which regulates epithelial to mesenchymal transition. PLoS 7(4), e35440 (2012)
22. Saeed, A.I., Bhagabati, N.K., Braisted, J.C., Liang, W., Sharov, V., Howe, E.A., et al.: TM4 microarray software suite. Meth. Enzymol. 411, 134–139 (2006)
23. Searson, D.P., Leahy, D.E., Willis, M.J., GPTIPS: an open source genetic programming toolbox for multigene symbolic regression. In: Proceedings of the International MultiConference of Engineers and Computer Scientists 2010 (IMECS 2010), Hong Kong, 17–19 March 2010
24. Bessarabova, M., Ishkin, A., JeBailey, L., Nikolskaya, T., Nikolsky, Y.: Knowledge-based analysis of proteomics data. BMC Bioinform. 13(16), S13 (2012)
25. Arsene, C.T.C., Lisboa, P.J.G., Borrachi, P., Biganzoli, E., Aung, M.S.H.: Bayesian neural networks for competing risks with covariates. In: Third International Conference in Advances in Medical, Signal and Information Processing, MEDSIP 2006, UK. IET (2006)
26. Arsene, C.T., Lisboa, P.J., Biganzoli, E.: Model selection with PLANN-CR-ARD. In: Cabestany, J., Rojas, I., Joya, G. (eds.) IWANN 2011, Part II. LNCS, vol. 6692, pp. 210–219. Springer, Heidelberg (2011)
27. Lisboa, P.J.L., Etchells, T., Jarman, I., Arsene, C.T.C., Aung, M.S.H., Eleuteri, A., Taktak, A.F.G., Ambrogi, F., Boracchi, P., Biganzoli, E.: Partial logistic artificial neural network for competing risks regularized with automatic relevance determination. IEEE Trans. Neural Netw. 20(9), 1403–1416 (2009)

# Integration of Clinico-Pathological and microRNA Data for Intelligent Breast Cancer Relapse Prediction Systems

Adriana Birlutiu[1,2], Denisa Ardevan[1], Paul Bulzu[1], Camelia Pintea[1,3(✉)], and Alexandru Floares[1]

[1] SAIA & OncoPredict, Cluj-Napoca, Romania
{adriana.birlutiu,denisa.ardevan,paul.bulzu,camelia.pintea,
alexandru.floares}@saia-institute.org
[2] "1 Decembrie 1918" University, Alba-Iulia, Romania
[3] Tech Univ Cluj-Napoca, North Univ Center Baia Mare, Baia Mare, Romania

**Abstract.** This paper investigates the integration of clinico-pathological and microRNA data for breast cancer relapse prediction. Clinical and pathological data proved to be relevant in making predictions about cancer disease outcome. The most accurate predictive models can be obtained by using clinico-pathological information together with genomic information. We analyzed the performance of various combinations between twenty classification algorithms and thirteen feature selection methods. The best performer was the regularized regression method Elastic Net, using its built-in feature selection method, on the data set integrating clinico-pathological data with microRNAs. The hybrid signature contains four clinico-pathological features and fifteen microRNAs. We also evaluated the influence of the separation of patients according to ER status and the impact of the exclusion from the data set of HS molecules (novel microRNAs without an assigned miRBase ID) on the overall performance. Functional analysis of the microRNAs of the best classifier showed that they are involved in cancer related processes.

**Keywords:** microRNA · Clinico-pathological data · Breast cancer relapse · Predictive models · Regularization

## 1 Introduction

Integrative analysis, mining, and modeling of various types of data are necessary prerequisites in the development of intelligent clinical decision support systems (i-CDSS) that would facilitate implementation of personalized medicine. Personalized medicine essentially involves the use of patient's omics data together with

This project has been conducted through the program Partnerships in priority areas - PN II, developed with the support of ANCS, CNDI - UEFISCDI, project no. PN-II-PT-CACM-2011-3.1-1221.

© Springer International Publishing Switzerland 2014
E. Formenti et al. (Eds.): CIBB 2013, LNBI 8452, pp. 178–193, 2014.
DOI: 10.1007/978-3-319-09042-9_13

clinical, pathological, and imaging information, to select a therapy or implement preventive measures that are particularly suited to certain patients. Integrating various types of data represents a key aspect in this context and can assist in the discovery of hybrid cancer signatures, while shedding new light on the molecular mechanisms underlying pathogenesis [15]. Despite significant efforts, cancer is still a lethal disease with a high mortality rate. Breast cancer occurs predominantly among women, being one the most common cancers. Worldwide, 1.4 million newly diagnoses and 460,000 breast cancer deaths are being reported each year [31]. Cancer research is currently focused on two major themes: the exploitation of biomarker panels for the development of clinical applications and a better understanding of the molecular mechanisms involved in carcinogenesis and cancer progression, at pathway-level.

Classical clinico-pathological markers incorporate experiences and conclusions accumulated over the years, and are usually relevant in making predictions about disease outcome. microRNAs are small, non-coding RNAs, mainly involved in the negative regulation of gene expression at the post-transcriptional and translational levels. They were identified as regulators in various types of disease, including breast cancer [23]. microRNAs were found differentially expressed between normal and tumor tissues and therefore, they can be used as disease predictors in different human cancers.

The current work analyzed the use of clinico-pathological and genomic data for developing intelligent systems for relapse prediction in breast cancer. We consider a binary classification problem: the response variable to be predicted can take two values, corresponding to disease relapse and no-relapse. We analyzed the predictive power of clinico-pathological markers and microRNAs, either alone or taken together. We tested various combinations between twenty classification algorithms and thirteen feature selection methods. The most accurate predictive models were obtained by integrating clinico-pathological and genomic data. The best accuracy (AUC = 0.87) was obtained with the regularized regression method Elastic Net, using its built-in feature selection. Elastic Net discovered a hybrid signature, integrating four clinico-pathological variables and fifteen microRNAs. The functional analysis of the selected microRNAs showed that they are related with various cancer processes. Thus, accuracy is increased and the cost is decreased by integrating classical clinico-pathological and microRNAs markers.

The paper is structured as follows. This section finishes with related works. The methods, including the experimental framework are presented in Sect. 2. The experimental evaluation and the biological interpretation of the results are presented in Sects. 3 and 4. Several conclusions are outlined in the last section.

## Related Works

Integrating clinico-pathological information was reported to improve performance estimated on expression data only in breast cancer [22,42]. In addition, other studies [12,28] enforce the position of clinico-pathological markers as powerful prognostic tools that overshadow the predictive capability of microarray expression profiles. Identifying a hybrid signature through the combination of genomic and clinico-pathological markers has been investigated in [39]. Classification performance of the hybrid signature on a survival data set outperforms the results obtained with a 70-gene signature of [41], with clinico-pathological makers alone, and with the St. Gallen consensus criterion [21]. Prognosis prediction for patients with breast cancer by integrating clinico-pathological and microarray data with Bayesian networks was studied by Gevaert et al. [20]. In the same field of interest, we previously investigated the development of transparent and interpretable intelligent systems for cancer diagnosis using decision tree models [16].

## 2   Methods

We consider a binary classification problem in which the response variable corresponds to breast cancer relapse and no-relapse. We analyze the predictive power of clinico-pathological markers and microRNAs, combined and separately.

### 2.1   Data Set

We used the breast cancer data set published in [6] and available on GEO [13] with the accession number: GSE22220. These data consist of expression profiling of mRNA and microRNA from 210 patients. Two patients were omitted from the original data set, as they have been lost sight of and thus the final data set comprised 208 case. Each patient's expression profile consisting of 735 microRNA and 24332 mRNA have been assessed. For the current study, we will use only the microRNA data, since evidence suggests a regulatory role for microRNAs in cancer [8,24]. The microRNA identifiers used here are of two types: *(i)* those with assigned miRBase IDs; and *(ii)* relatively novel microRNAs that are identified as HS sequences. Each patient in the data set has associated information regarding disease relapse. Disease relapse in cancer is defined as the recurrence of an identical tumor after complete cure in a specific time interval, which is ten years for the data set used here.

In addition to gene expression data, clinico-pathological information has been collected for every patient. The clinico-pathological information consists of five patient characteristics: age, tumor size, tumor grade, nodes involved, and ER (estrogen receptor) status. According to Eifel et al. [14], clinico-pathological data can be classified in: *(i)* patient characteristics that are independent of the disease; *(ii)* disease characteristics, for example the tumor size; and *(iii)* measurable parameters in tissue cells or fluids, such as estrogen receptor status.

1. *Patient age* has the average of 55 years with a standard deviation of 10 years.
2. *Tumor size* is a variable with values ranging from 0 to 7.
3. *Tumor grade* is a variable used to classify tumor cells into 3 categories according to their differentiation status.
4. *Nodes involved* is a clinical variable ranging from 0 to 16.

5. *Estrogen receptor (ER) status* is a binary variable used to distinguish between two major tumor groups: ER−positive and ER−negative. These tumor groups differ with respect to endocrine sensitivity to estrogen or estrogen-mimicking substances.

## 2.2   Experimental Framework

The data set was split into training and testing in a cross-validation procedure with 10 folds. On each fold, training data set was used for learning the model, feature selection and classifier, and testing data was used for evaluating the learned model.

We compared twenty classification algorithms[1] in combination with thirteen feature selection methods[2] to decide on which to further focus for tunning. We took as input: microRNA data only, clinico-pathological data only, and their combination. The experimental evaluation was performed in R language [34], using at first the *CMA* package. The *CMA* [38] package provides an interface that allows the user to easily run and compare a large set of classification algorithms. The algorithms were run with their default parameters and performance measure was considered in terms of misclassification error.

Out of all evaluated classification algorithms, we selected six that showed the best performance: Componentwise Boosting, Elastic Net, Lasso, Penalized Logistic Regression, Random Forests and Support Vector Machines. Componentwise Boosting combines linear functions in an ensemble; it generates sparsity and can be as used for variable selection alone [5]. Penalized Logistic Regression adds an L2-type penalty to the high dimensional logistic regression [49]. Random Forests [4] is a learning ensemble that constructs decision tree learners and randomly selects features at each split. SVM [37] is able to perform both linear and non-linear classifications by constructing one or several hyperplanes in a high dimensional space. Performance of these selected algorithms was further tested over different values of selected variables. The algorithm with the best result was chosen for further analysis.

Linear models have regained popularity in machine learning applications for big datasets where the number of features is much larger than the number of samples. In this context, our focus on Elastic Net enforces this observation. Elastic Net [50] is a regularized regression method combining the L1 and L2 penalties of

---

[1] Componentwise Boosting, Diagonal Discriminant Analysis, Elastic Net, Fisher Discriminant Analysis, Tree-based Boosting, k-nearest neighbors, Linear Discriminant Analysis, Lasso, Feed-Forward Neural Networks, Probabilistic nearest neighbors, Penalized Logistic Regression, Partial Least Squares with Linear Discriminant Analysis, Partial Least Squares with logistic regression, Partial Least Squares with Random Forest, Probabilistic Neural Networks, Quadratic Discriminant Analysis, Random Forest, PAM, Shrinkage Discriminant Analysis, Support Vector Machine.

[2] t test, Welch test, Wilcox test, F test, Kruskal-Wallis test, moderated t and F test (limma), One-step Recursive Feature Elimination, random forest variable importance measure, Lasso, Elastic Net, componentwise boosting, Golub ad-hoc criterium, shrinkcat.

the Lasso and Ridge methods. This method encourages a grouping effect, where strongly correlated predictors tend to be in or out of the model together. Elastic Net is particularly suitable in classification problems on data where the number of predictors is much larger than the number of observations. Microarray data constitutes a classical example: thousands of transcripts measured, while patients are at maximum a few hundreds.

The classification problem can be formalized as follows: we have a predictor space $\mathcal{X} \in \mathbb{R}^p$. We are given a training set with $n$ observations $\{(x_1, y_1), \ldots, (x_n, y_n)\}$ where $x_i$ denotes a $p$ dimensional predictor vector and $y$ is a binary response variable encoding disease relapse or no-relapse. The task is to find a decision function:

$$\hat{f} : \mathcal{X} \rightarrow \{0, 1\}$$
$$x \rightarrow \hat{f}(x)$$

where $\hat{}$ means that the prediction function is being estimated from the training sample such that the generalization error is minimized. For linear models, this error is defined as:

$$\mathcal{L}[(f(x), y)] = \sum_{i=1}^{n} (y_i - x_i^T \beta)^2$$

In order to reduce overfitting, a penalty term is added to the loss function:

$$\mathcal{L}[(f(x), y)] = \sum_{i=1}^{n} (y_i - x_i^T \beta)^2 + \lambda P_\alpha(\beta) \tag{1}$$

Ridge regression ($\alpha = 1$), Elastic Net ($0 < \alpha < 1$) and Lasso ($\alpha = 0$) are part of the same family with general form of the penalty term:

$$P_\alpha(\beta) = \sum_{j=1}^{p} \left[ \frac{1}{2}(1-\alpha)\beta_j^2 + \alpha|\beta_j| \right] \tag{2}$$

The experimental evaluation was performed using the *glmnet* R package [18]. This package is a fast implementation of the standard coordinate descent algorithm for solving $\mathcal{L}_1$ penalized learning problems. Models are built in a so-called coordinate descent: optimization of each parameter is done separately, maintaining all other parameters fixed and $\lambda$ values in a descending scale.

## 3    Results

We first analyzed the relation between two clinical variables and disease relapse. The clinical variables that we considered were: tumor size and nodes involved. We observed that tumor size is related to disease relapse, and, in general, the larger the tumor, the higher the chances of disease relapse. This can be seen in the plot shown on the lefthand side of Fig. 1: for small tumor sizes, 1 or 2,

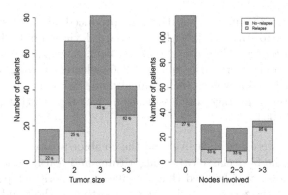

**Fig. 1.** Left: Histogram of patient number versus tumor size. Right: Histogram of patient number versus number of nodes involved. The color of the bars in the histograms show the division of patients in the two groups: relapse (light gray) and no-relapse (dark gray). The numbers show the percentages of patients with relapse.

the percentage of relapse is between 20–25 %, while for tumor sizes larger than 3, the percentage reaches 60 %. A certain correlation degree can be observed between the number of nodes involved and disease relapse by analyzing the plot from the righthand side of Fig. 1: for a small number of nodes, the risk of relapse is smaller than for a larger number of nodes involved.

We compared several classification algorithms in combination with several feature selection methods to decide on which to concentrate for further tunning. The comparison was performed using *CMA* package in R language.

1. *Clinical data only.* The classification algorithms were tested using clinical data only. The tumor grade clinical variable is missing for a number of 23 patients. To screen the effect over classification performance of the tumor grade variable three input sets were tested: first, the tumor grade variable was excluded from the input data set. Second, tumor grade missing values were imputed using the k-Nearest Neighbor algorithm (kNN) [11] available in the package imputation [45]. The optimal number of neighbors to use was determined by cross validation ($k = 4$) and passed on the imputation. Third, patients with missing tumor grade values were excluded from the data set. Tumor grade variable affected the average classification performance as follows. When it was excluded from the clinico-pathological variables, the average misclassification error was 0.321 and standard deviation of 0.056. Evaluation on input data that excluded patients with missing values for tumor grade resulted in 0.335 average misclassification error and standard deviation of 0.059. The best average performance, 0.318 and standard deviation of 0.050, was obtained when missing values were imputed with kNN. Further analyses were conducted using tumor grade imputed with kNN algorithm.
2. *Genomic data only.* Considering that some classification algorithms have built-in feature selection, the number of variables selected from the feature selection outcome was relatively high, specifically 250. Classification

algorithms tested on microRNA data only produced fair results: average misclassification error was 0.357, with a standard deviation of 0.014.

3. *Combination between clinical and genomic data.* As explained above, classification algorithms were also run on 250 selected variables. Combining microRNA expression values with clinical data, resulted in overall improved accuracy. Results: average misclassification error was 0.334 with a standard deviation of 0.054.

Further analyses were run exclusively on the combined data set.

Selection of the best classification algorithms and feature selection method was done as follows: we observed that classification algorithms performed best in combination with feature selection using Boosting method and set an upper threshold of 0.280, corresponding to the misclassification error obtained with Random Forest and Boosting. Feature selection methods that included at least three classification algorithms with performance below the threshold were maintained for the analysis and all the remaining combinations were further tested. Table 1 shows the selected algorithms with their combined feature selection, highlighted in a grey background. Best accuracy 0.267 was obtained with Componentwise Boosting classification algorithm and Shrinkcat as feature selection.

The selected algorithms and feature selection methods were run for a range of selected number of variables: $20, 50, 100, 150, \ldots, 300$, for the combination between clinical and microRNA data. Results are shown in Fig. 2. From the tested combinations of feature selection methods and classification algorithms, the smallest misclassification error was obtained with Shrinkcat as feature selection method and Elastic Net as classifcation algorithm.

Since Elastic Net has built-in feature selection, we further tested its performance using the implementation from *glmnet* package on the combination between clinical and microRNA data. Classification performance of Elastic Net was estimated using area under ROC curve (AUC) [46]. We considered different

**Table 1.** Comparison of classification algorithms (Componentwise Boosting, Elastic Net, Lasso, Penalized Logistic Regression, Random Forest and SVM) in combination with several feature selection methods. Performance was measured in terms of misclassification error.

|  | Boosting | Elastic Net | Kruskal-Wallis | Shrinkcat |
|---|---|---|---|---|
| Componentwise Boosting | 0.271 | 0.275 | 0.267 | 0.267 |
| Elastic Net | 0.350 | 0.276 | 0.268 | 0.272 |
| Lasso | 0.349 | 0.275 | 0.280 | 0.277 |
| Penalized Logistic Regression | 0.2690 | 0.298 | 0.417 | 0.305 |
| Random Forest | 0.280 | 0.289 | 0.311 | 0.316 |
| SVM | 0.275 | 0.312 | 0.333 | 0.303 |

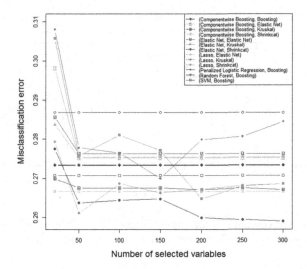

**Fig. 2.** Classification performance for all combinations of feature selection methods and classification algorithms. Misclassification error was used for measuring the performance.

parameter configurations for the values of $\alpha = \{0.1, 0.2, \ldots, 0.9\}$ (see Eqs. 1 and 2) using 10 fold cross-validation. Best AUC performance over the whole data set was 0.87, corresponding to $\alpha = 0.6$, $\lambda = 0.11$ and only *nineteen* selected variables. This AUC value is higher than the AUC obtained by Buffa et al. [6] (the AUC's obtained by Buffa et al. [6] using Cox models is around 0.71). The algorithm's selected variables were also screened during the process. For each $\alpha$ value tested, we observed that four out of the total of five clinico-pathological variables were always selected: *age, tumor size, nodes involved* and *ER status*. This aspect confirms their important role as predictors along with genomic data.

Basically, the classification algorithm developed will be used to make a prediction about cancer relapse for new patients, and this prediction will help in choosing the appropriate treatment for that patient (for example, patients with a high risk of relapse will be given a more aggressive treatment). By considering and screening only 15 microRNAs instead of about 700, the costs associated are considerably reduced.

For comparison purposes, we have investigated the results and the value of $\lambda$ by using a cross-validation procedure, for $\alpha = 0$ and $\alpha = 1$ corresponding to the Lasso algorithm and the Ridge regression (Fig. 3). The dotted lines indicate the most stable interval of $\lambda$ values corresponding to the best performances. It can be noticed that Ridge regression algorithm produces less robust results, with a larger standard deviation. Lasso classification algorithm shows a similar behavior to Elastic Net.

### 3.1 Patient Differentiation Based on ER Status

The ER status variable has the capability to distinguish between tumor cell populations of different origins and it is an important parameter used in practice

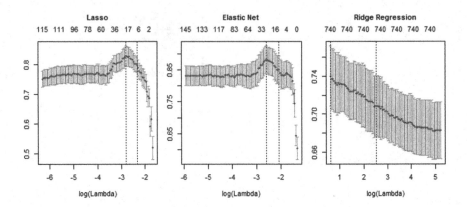

**Fig. 3.** Performance comparison for Lasso, Elastic Net and Ridge Regression for different values of $\lambda$ parameter. AUC was used as the evaluation metric.

by clinicians. This is why we wanted to evaluate two models of classifiers for ER+ patients, respectively ER−. The original group of patients was divided into two subsets based on the ER status. We obtained 81 patients with ER positive status and 127 patients with ER negative status. We investigated the performance of 18 classification algorithms and 11 selection methods for the two subsets of data obtained.

To assess whether the classification algorithms perform better after this differentiation, we used the t-test. We compared the performances for the full set of patients with the performances on the ER+ subset. On average, a better performance was obtained on the complete set for 12 classification algorithms, and for 8 algorithms this difference was significant. For the 5 classifiers that had a better accuracy on the data set with ER+ patients, the difference was significant only for two of them. Similarly, 11 classification algorithms had better average results on the complete set of patients compared with the performance of classifiers tested on ER− group of patients, the difference being significant for 7 of them. Of the remaining 6 algorithms, only 2 had a significantly better performance when running ER− data.

This separation would be worthy of further consideration if the results clearly indicated a significant increase in performance, which they do not, in this case.

## 3.2 Exclusion of HS Sequences

From the total of 735 micoRNA molecules, only 488 of them have an identifier in the miRBase [29]. We investigated whether the performance improves when the microRNA molecules which are not classified in the miRBase (the HS sequences) are excluded from the data set. To make this comparison we used the t-test. The analysis showed that 15 (out of 18 algorithms) had a better average accuracy in classifying all 735 microRNA molecules, the difference being significant for 8 of them. This suggests that the HS molecules are involved in cancer recurrence.

# 4    Biological Significance

Among the fifteen objects from the initial list, only ten were confirmed as microR-
NAs with assigned miRBase IDs, and we consider these ten microRNAs as the list
of informative microRNAs. The ten microRNAs found to be informative regarding disease relapse are: *hsa-miR-668, hsa-miR-566, hsa-miR-487a, hsa-miR-451,
hsa-miR-448, hsa-miR-199a-5p, hsa-miR-191, hsa-miR-18a, hsa-miR-106a:9.1,
hsa-miR-1260*. None of these microRNAs matched any of those determined in the
original study conducted by Buffa et al. [6]. We further subjected the microRNAs
to functional analysis, thus bringing them into biologically meaningful context.
The functional role of each informative microRNA was investigated by consulting the available PubMed literature on similar studies. Integrated analysis was
then conducted for all informative microRNAs using three web-based enrichment
analysis applications: MetaCore$^{TM}$, DAVID v6.7 [27] and ToppGene [10]. We
further discuss our findings.

Expression levels of hsa-miR-106a:9.1 have been previously investigated by
Wang et al. [44] in both tumor and adjacent normal tissues, as well as in
serum derived from blood samples. Results have indicated the overexpression
of this microRNA in the tumor tissue samples and in the serum of breast cancer
patients. One known role of hsa-miR-106a:9.1 is the inhibition of macrophage
development, thus limiting the immune response towards cancer cells [17]. The
overexpression of hsa-miR-18a in breast cancer cell lines and tumor tissue was
linked to the downregulation of ATM kinase which is normally involved in DNA
damage detection. This study has confirmed ATM as being a target of hsa-miR-
18a, whose increased expression levels lead to lowered DNA repair capabilities.
Also, significantly higher levels of hsa-miR-18a where observed in ER+ tumors
than in ER− ones [9]. Volinia et al. have shown that hsa-miR-191, together
with hsa-miR-21 and hsa-miR-17-5p, are expressed at higher levels in all the
types of cancer that they have studied [43]. Evidence shows that hsa-miR-191
is involved in the epithelial to mesenchymal transition (EMT) [26]. Also, the
role of hsa-miR-448 in EMT has been highlighted by Li et al. [30] using breast
cancer cell lines. The involvement of hsa-miR-199a-5p in autophagy has been
documented in a study conducted by Yi et al. [47]. Even though autophagy has
been associated with various diseases, including cancer, the exact mechanisms
involved have not been elucidated since. The overexpression of hsa-miR-199a-
5p inhibited radiation induced autophagy in a breast cancer cell line. This is
achieved by repression of two key proteins involved in autophagy activation:
DRAM1 and Beclin1, by hsa-miR-199-5p. In a recent study, Bergamaschi et al.
[3] have demonstrated the link between hsa-miR-451 and breast cancer recurrence and another study has highlighted the relationship between this microRNA
and resistance to cancer treatment [48]. Seven microRNAs, together with hsa-
miR-487a, from the 14q32 chromosomal region have shown lowered expression
levels in Ewig sarcoma metastatic xenografts, in comparison to control samples,
thus suggesting a potential role in tumor suppression for these microRNAs [33].

No relevant literature was found linking hsa-miR-668, -566 or -1260 to breast
cancer. However, we have used a freely accessible web-based tool called

MIRUMIR [2] to perform survival analyses for submitted microRNA IDs across multiple available GEO data sets. MIRUMIR makes use of rank information from these data sets. For each set, samples are grouped according to the expression rank of the miRNA that is specified by the user. The two groups : "low expression" and "high expression", are those in which the expression rank of the microRNA is lower or higher than the average expression rank across the whole data set. In order to find statistical differences in patient survival outcome, this separation into low and high groups is used along with survival information [2]. Nine out of ten microRNAs from our list were thus found to be significantly linked to poor survival outcome in breast cancer when expressed at low levels. In this case, data set GSE37405 was analysed for microRNA ranking. No expression data was available for hsa-miR-1260 however.

Functional analysis for the ten microRNAs was first performed in MetaCore$^{TM}$ using Dijkstra's shortest paths algorithm to add known targets and intermediary molecules linking the microRNAs in an expanded network. Adequate connectivity was reached when choosing four as the maximum number of steps in the path between objects. We exported the gene list associated with the obtained network and analyzed it using the functional enrichment analysis application in MetaCore$^{TM}$ software [1]. Figure 4 shows the top ten pathway maps (left plot) and processes (right plot) identified using our initial microRNAs list together with their targets and intermediary molecules. The pathway maps and processes are ordered according to their statistic significance (log(pValue)). The first ranking pathway map that was identified points out to cytoskeleton remodeling processes together with Wnt and TGF-$\beta$ signaling, all of which are relevant to cancer progression, tumor growth, differentiation and metastasis [7,32]. TGF-$\beta$ also modulates processes such as cell invasion, immune regulation and tumor microenvironment modification [32]. The activation of IGF-1 receptor signaling

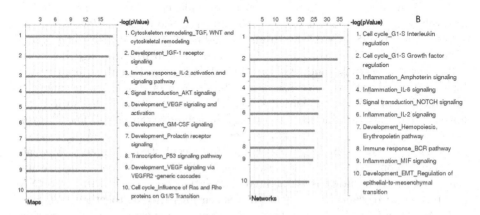

**Fig. 4.** The top ten pathway maps (left plot) and processes (right plot) identified using our initial microRNAs list together with their targets and intermediary molecules. The analysis was performed using $MetaCore^{TM}$ software [1]. The pathway maps and processes are ordered according to their statistic significance ($-\log(\text{pValue})$).

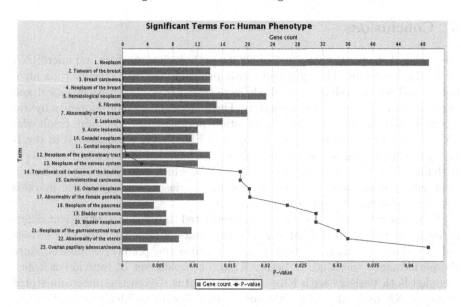

**Fig. 5.** The highest ranking phenotypes associated with the informative microRNA list and their targets and intermediary molecules identified using ToppGene [10]. The top scale of the graph is concordant with gene count and the bottom scale with P-value.

was identified in the second highest ranking pathway map, and it is known to be highly correlated to disease progression, radiotherapy resistance and a generally unfavorable prognostic in breast cancer [35,40]. Interleukin 2 (IL-2) activation and signaling was also identified as a relevant process for our list of microR-NAs. The IL-2 receptors activate several different pathways that mediate the flow of mitogenic and survival-promoting signals [7] but also play a critical role in orchestrating the immune responses [19]. Another relevant map points out to the process of angiogenesis, induced by VEGF signaling, which is required for invasive tumor growth and metastasis [36].

In summary, all of the top ten pathway maps and processes that have been identified by us are known as key players in carcinogenesis and immune suppression [25]. Functional annotation of the extended list obtained in MetaCore TM was analyzed in DAVID v6.7 [27] (The Database for Annotation, Visualization and Integrated Discovery) and returned the top KEGG Pathway term Pathways in cancer, with a pValue of 3.0E-42. ToppGene [10] identified the highest ranking phenotypes associated with these genes as neoplasms and breast carcinoma, both with pValues <<0.005 (Fig. 5).

MicroRNAs associated with distant relapse-free survival identified by Buffa et al. [6] (hsa-miR-29c, hsa-miR-642, hsa-miR-548d), considering both ER+ and ER- patients, returned similar results: top KEGG Pathway term Pathways in cancer, with a pValue of 3.8E-8 and neoplasms as highest ranking phenotype in ToppGene, with a pValue below 0.015.

## 5 Conclusions

We analyzed the predictive power of clinico-pathological markers and microRNA in the disease relapse binary classification problem. Multiple classification algorithms and feature selection methods were tested over clinico-pathological and genomic data, combined and separately. The focus was set on developing a hybrid signature that would improve classification performance. The best result was obtained with regularized regression method Elastic Net implemented in the R package *glmnet*. The marker variables selected included four out of five clinico-pathological data and fifteen other microRNAs which were found to be associated with various cancer processes. Integrating classical and genomic markers can increase accuracy and decrease cost.

Using the integrated data, we also investigated the efficiency of patient differentiation according to the ER status. The results obtained for the two subsets did not show that this strategy would optimize the predictive models, since the performance was slightly better for the complete set of individuals which included both patients with ER+ and ER− status. Given the uncertain status of microRNA molecules which are not associated with an identifier in the miR-Base (HS sequences), they were excluded from the predictor variables. Testing the combinations of classification algorithms and feature selection methods on the data set with HS sequences excluded showed slightly lower performance, suggesting that HS molecules are involved in breast cancer recurrence.

## References

1. Metacore gene expression and pathway analysis. http://www.genego.com/metacore.php
2. Antonov, A.V., Knight, R.A., Melino, G., Barlev, N.A., Tsvetkov, P.O.: Mirumir: an online tool to test micrornas as biomarkers to predict survival in cancer using multiple clinical data sets. Cell Death Differ. **20**(2), 367 (2013). http://dx.doi.org/10.1038/cdd.2012.137L3
3. Bergamaschi, A., Katzenellenbogen, B.S.: Tamoxifen downregulation of mir-451 increases 14-3-3zeta and promotes breast cancer cell survival and endocrine resistance. Oncogene **31**(1), 39–47 (2012)
4. Breiman, L.: Random forests. Mach. Learn. **45**(1), 5–32 (2001)
5. Buelmann, P., Yu, B.: Boosting with the l2 loss: regression and classification. J. Am. Stat. Assoc. **98**, 324–339 (2003)
6. Buffa, F.M., Camps, C., Winchester, L., Snell, C.E., Gee, H.E., Sheldon, H., Taylor, M., Harris, A.L., Ragoussis, J.: Microrna-associated progression pathways and potential therapeutic targets identified by integrated mrna and microrna expression profiling in breast cancer. Cancer Res. **71**(17), 5635–5645 (2011)
7. Burns, L.J., Weisdorf, D.J., et al.: Il-2-based immunotherapy after autologous transplantation for lymphoma and breast cancer induces immune activation and cytokine release: a phase i/ii trial. Bone Marrow Transplant. **32**(2), 177–186 (2003)
8. Calin, G.A., Croce, C.M.: MicroRNA signatures in human cancers. Nat. Rev. Cancer **6**(11), 857–866 (2006)

9. Castellano, L., Giamas, G., et al.: The estrogen receptor-alpha-induced microrna signature regulates itself and its transcriptional response. Proc. Natl. Acad. Sci. USA **106**(37), 15732–15737 (2009)
10. Chen, J., Bardes, E., Aronow, B., Jegga, A.: Toppgene suite for gene list enrichment analysis and candidate gene prioritization. Nucleic Acids Res. **37**(suppl 2), W305–W311 (2009)
11. Duda, R., Hart, P., Stork, D.: Pattern Classification. Wiley-Interscience, New-York (2001)
12. Edén, P., Ritz, C., Rose, C., Fernö, M., Peterson, C.: "Good old" clinical markers have similar power in breast cancer prognosis as microarray gene expression profilers. Eur. J. Cancer **40**, 1837–1841 (2004)
13. Edgar, R., Domrachev, M., Lash, A.E.: Gene expression omnibus: NCBI gene expression and hybridization array data repository. Nucleic Acids Res. **30**(1), 207–210 (2002)
14. Eifel, P., Axelson, J.A., Costa, J., Crowley, J., Curran, W.J., Deshler, A., Fulton, S., Hendricks, C.B., Kemeny, M., Kornblith, A.B., Louis, T.A., Markman, M., Mayer, R., Roter, D.: National institutes of health consensus development conference statement: adjuvant therapy for breast cancer, November 1–3, 2000. J. natl. cancer inst. **93**(13), 979–989 (2001)
15. Famili, F., Phan, S., Fauteux, F., Liu, Z., Pan, Y.: Data integration and knowledge discovery in life sciences. In: García-Pedrajas, N., Herrera, F., Fyfe, C., Benítez, J.M., Ali, M. (eds.) IEA/AIE 2010, Part III. LNCS (LNAI), vol. 6098, pp. 102–111. Springer, Heidelberg (2010)
16. Floares, A., Birlutiu, A.: Decision tree models for developing molecular classifiers for cancer diagnosis. In: The 2012 International Joint Conference on Neural Networks (IJCNN), pp. 1–7. IEEE (2012)
17. Fontana, L., Pelosi, E. et al.: MicroRNAs 17–5p-20a-106a control monocytopoiesis through AML1 targeting and M-CSF receptor upregulation. Nat. Cell Biol. **9**(7), 775–787 (2007). http://dx.doi.org/10.1038/ncb1613
18. Friedman, J., Trevor, H., Tibshirani, R.: Regularization paths for generalized linear models via coordinate descent. J. Stat. Softw. **33**(1), 1–22 (2010). http://www.jstatsoft.org/v33/i01/
19. Gaffen, S.L., Liu, K.D.: Overview of interleukin-2 function, production and clinical applications. Cytokine **28**(3), 109–123 (2004). http://www.sciencedirect.com/science/article/pii/S1043466604002200
20. Gevaert, O., Smet, F.D., Timmerman, D., Moreau, Y., Moor, B.D.: Predicting the prognosis of breast cancer by integrating clinical and microarray data with bayesian networks. Bioinformatics **22**(14), e184–e190 (2006)
21. Goldhirsch, A., Wood, W.C., Gelber, R.D., Coates, A.S., Thürlimann, B., Senn, H.J.: Meeting highlights: updated international expert consensus on the primary therapy of early breast cancer. J. Clin. Oncol. **21**(17), 3357–3365 (2003)
22. González, S., Guerra, L., Robles, V., Peña, J., Famili, F.: Clidapa: a new approach to combining clinical data with dna microarrays. Intell. Data Anal. **14**(2), 207–223 (2010)
23. Guo, L., Zhao, Y., Yang, S., Cai, M., Wu, Q., Chen, F.: Genome-wide screen for aberrantly expressed mirnas reveals mirna profile signature in breast cancer. Mol. Biol. Rep. **40**(3), 2175–2186 (2013)
24. Han, Y., Chen, J., et al.: MicroRNA expression signatures of bladder cancer revealed by deep sequencing. PLoS ONE **6**(3), 6 (2011)
25. Hanahan, D., Weinberg, R.: Hallmarks of cancer: the next generation. Cell **144**(5), 646–674 (2011)

26. He, Y., Cui, Y., et al.: Hypomethylation of the hsa-mir-191 locus causes high expression of hsa-miR-191 and promotes the epithelial-to-mesenchymal transition in hepatocellular carcinoma. Neoplasia **13**(9), 841–853 (2011)

27. da Huang, W., Sherman, B., Lempicki, R.: Systematic and integrative analysis of large gene lists using david bioinformatics resources. Nature Protoc. **1**, 44–57 (2008)

28. Ioannidis, J.P.: Microarrays and molecular research: noise discovery? Lancet **365**(9458), 454–455 (2005)

29. Kozomara, A., Griffiths-Jones, S.: miRBase: integrating microRNAannotation and deep-sequencing data. Nucleic Acids Res. 39(Database-Issue), 152–157 (2011). http://dblp.uni-trier.de/db/journals/nar/nar39.html#KozomaraG11d

30. Li, Q.Q., Chen, Z.Q., et al.: Involvement of NF-kappaB/miR-448 regulatory feedback loop in chemotherapy-induced epithelial-mesenchymal transition of breast cancer cells. Cell Death Differ. **18**(1), 16–25 (2011)

31. Ma, J., Jemal, A.: Breast cancer statistics. In: Ahmad, A. (ed.) Breast Cancer Metastasis and Drug Resistance, pp. 1–18. Springer, New York (2013)

32. Massague, J.: TGFbeta in cancer. Cell **134**(2), 215–230 (2008)

33. Mosakhani, N., Guled, M., et al.: An integrated analysis of miRNA and gene copy numbers in xenografts of Ewing's sarcoma. J. Exp. Clin. Cancer Res. **31**, 24 (2012)

34. R Development Core Team: R: A language and environment for statistical computing. R Foundation for Statistical Computing 1(2.11.1), 409 (2011). http://www.r-project.org

35. Rocha, R.L., Hilsenbeck, S.G., et al.: Insulin-like growth factor binding protein-3 and insulin receptor substrate-1 in breast cancer: correlation with clinical parameters and disease-free survival. Clin. Cancer Res. **3**(1), 103–109 (1997)

36. Schoeffner, D.J., Matheny, S.L., et al.: VEGF contributes to mammary tumor growth in transgenic mice through paracrine and autocrine mechanisms. Lab Invest. **85**(5), 608–623 (2005)

37. Schölkopf, B., Smola, A.J.: Learning with Kernels: Support Vector Machines, Regularization, Optimization, and Beyond. MIT Press, Cambridge (2002). http://www.learning-with-kernels.org

38. Slawski, M., Boulesteix, A.L., Bernau., C.: CMA: Synthesis of microarray-based classification, r package version 1.16.0. (2009)

39. Sun, Y., Goodison, S., Li, J., Liu, L., Farmerie, W.: Improved breast cancer prognosis through the combination of clinical and genetic markers. Bioinformatics **23**(1), 30–37 (2007)

40. Turner, B.C., Haffty, B.G., et al.: Insulin-like growth factor-I receptor overexpression mediates cellular radioresistance and local breast cancer recurrence after lumpectomy and radiation. Cancer Res. **57**(15), 3079–3083 (1997)

41. van't Veer, L.J., Dai, H., Van De Vijver, M.J., He, Y.D., Hart, A.A., Mao, M., Peterse, H.L., van der Kooy, K., Marton, M.J., Witteveen, A.T., et al.: Gene expression profiling predicts clinical outcome of breast cancer. Nature **415**(6871), 530–536 (2002)

42. van Vliet, M.H., Horlings, H.M., van de Vijver, M.J., Reinders, M.J., Wessels, L.F.: Integration of clinical and gene expression data has a synergetic effect on predicting breast cancer outcome. PLoS ONE **7**(7), e40358 (2012)

43. Volinia, S., Calin, G.A., et al.: A microRNA expression signature of human solid tumors defines cancer gene targets. Proc. Natl. Acad. Sci. USA **103**(7), 2257–2261 (2006)

44. Wang, F., Zheng, Z., Guo, J., Ding, X.: Correlation and quantitation of microRNA aberrant expression in tissues and sera from patients with breast tumor. Gynecol. Oncol. **119**(3), 586–593 (2010)
45. Wong, J.: Package 'imputation', version 2.0.1. https://github.com/jeffwong/imputation
46. Xiao-Hua, Z., Obuchowski, N., McClish, D.: Statistical methods in diagnostic medicine (2002)
47. Yi, H., Liang, B., et al.: Differential roles of miR-199a-5p in radiation-induced autophagy in breast cancer cells. FEBS Lett. **587**(5), 436–443 (2013)
48. Zhu, H., Wu, H., Liu, X., Evans, B.R., Medina, D.J., Liu, C.G., Yang, J.M.: Role of microRNA miR-27a and miR-451 in the regulation of MDR1/P-glycoprotein expression in human cancer cells. Biochem. Pharmacol. **76**(5), 582–588 (2008)
49. Zhu, J., Hastie, T.: Classification of gene microarrays by penalized logistic regression. Biostatistics **5**(3), 427–443 (2004)
50. Zou, H., Hastie, T.: Regularization and variable selection via the elastic net. J. Roy. Stat. Soc.: Ser. B (Stat. Methodol.) **67**(2), 301–320 (2005)

# Superresolution MUSIC Based on Marčenko-Pastur Limit Distribution Reduces Uncertainty and Improves DNA Gene Expression-Based Microarray Classification

Leif E. Peterson[✉]

Center for Biostatistics, The Methodist Hospital Research Institute,
Weill Cornell Medical College, Cornell University, 6565 Fannin,
Suite MGJ6-031, Houston, TX 77030, USA
lepeterson@houstonmethodist.org

**Abstract.** We introduce a bootstrap root MUSIC (BRM) technique, which employs superresolution multisignal classification to reduce high-dimensional sets of genes from expression microarrays to low-dimensional sets used in supervised classification analysis. During BRM, the Marčenko-Pastur limit distribution of eigenvalues for the array-by-array gene expression covariance matrix was used for determining the eigenvalue cutoff for the noise subspace. Classifier results were compared with and without replacing gene expression values with the inverse of the distance to class-specific noise eigenspace for each microarray. Nine gene expression datasets were used for classification, and results of using BRM were compared with classification results based on use of random and best ranked N genes. On average, BRM resulted in greater classification of randomly selected genes when compared with direct use of randomly selected genes for classifier input. In addition, when BRM was applied to best ranked N genes, the interquartile ranges of accuracy were smaller when compared with direct input of best ranked genes into classifiers. Overall, BRM can optimally be used with 128 or 256 best ranked markers, requiring less extensive filtering to identify smaller sets of predictors. Use of a larger set of markers with BRM can help minimize the effect of concept drift over time.

## 1 Introduction

Supervised classification analysis of DNA microarrays based on expression of messenger RNA, oligonucleotides, or micro RNAs in various disease or exposure classes is a common enterprise in etiologic research. Under most circumstances, an "optimal" gene set is identified via filtering, where selection of the best predictors is done independent of the classifier used, or by wrapping, where the classifier itself identifies the best predictors. "Best" is taken to mean either a statistically significant set of joint predictors identified simultaneously or the best ranked N predictors identified singly. Wrapping is known to suffer from a

© Springer International Publishing Switzerland 2014
E. Formenti et al. (Eds.): CIBB 2013, LNBI 8452, pp. 194–209, 2014.
DOI: 10.1007/978-3-319-09042-9_14

varying degree of selection bias because the genes identified can make up for weaknesses of the classifier employed [1]. For this reason, there is typically less selection bias when a wholly independent filtering-based approach is initially used, followed by classifier training and testing.

The informativeness of an optimal set of gene predictors is inextricably hinged to the bias-variance relationship for the data and model used [2]. A classifier model having too many features (degrees of freedom) will likely result in over-fitting, where system noise is also learned. In such cases, the bias will be consistently low implying a good model fit; however, the amount of variance will be moderate as the high fitness levels will change with the data. Analogously, when there are too few degrees of freedom, the classifier will usually fit most data sets poorly resulting in a consistently high level of bias – in this case the data used has a lesser impact. The point is clear: overfitted models with too many degrees of freedom are more sensitive to the data (high variance of fitness across datasets), whereas poorly fitting under-parametrized models are characteristically less sensitive to the data and therefore inaccurate (high bias across datasets).

An alternative solution for the bias-variance dilemma is to employ superresolution multiple signal classification (MUSIC), which has been shown to provide significant advantages in high-dimensional RADAR recognition involving range, azimuth, and target identification [3]. Superresolution MUSIC provides asymptotically unbiased estimates of RADAR target characteristics and has been shown to outperform beamforming, maximum likelihood, and maximum entropy techniques [4]. The fundamental benefit of superresolution MUSIC is that it exploits the orthogonality between the eigenvector noise subspace of the feature-by-feature covariance matrix and the direction vector for each unknown target [5]. In this study, we investigate the performance of an ensemble of classifiers employed for class prediction of 9 gene expression datasets with and without the use of superresolution MUSIC.

## 2   Methods

### 2.1   Datasets Used

A 2-class adult brain cancer dataset was comprised of 60 arrays (21 censored, 39 failures) with expression for 7,129 genes [6]. A 2-class adult prostate cancer data set consisted of 102 training samples (52 tumor, and 50 normal) with 12,600 features. The original report for the prostate data supplement was published by Singh et al. [7]. Two breast cancer data sets were used. The first had 2 classes and consisted of 15 arrays for 8 BRCA1 positive women and 7 BRCA2 positive women with expression profiles of 3,170 genes [8], and the second was also a 2-class set including 78 patient samples and 24,481 features (genes) comprised of 34 cases with distant metastases who relapsed ("relapse") within 5 years after initial diagnosis and 44 disease-free ("non-relapse") for more than 5 years after diagnosisl [9]. Two-class expression data for adult colon cancer were based on the paper published by Alon et al. [10]. The data set contains 62 samples based on

**Table 1.** Data sets used for classification analysis.

| Cancer site | Classes | Samples | Features | Reference |
|---|---|---|---|---|
| Brain | 2 | 60 (21 censored, 39 failures) | 7,129 | Pomeroy et al. [6] |
| Prostate | 2 | 102 (52 tumor, 50 normal) | 12,600 | Singh et al. [7] |
| Breast | 2 | 15 (8 BRCA1, 7 BRCA2) | 3,170 | Hedenfalk et al. [8] |
| Breast | 2 | 78 (34 relapse, 44 non-relapse) | 24,481 | van 't Veer et al. [9] |
| Colon | 2 | 62 (40 negative, 22 positive) | 2,000 | Alon et al. [10] |
| Lung | 2 | 32 (16 MPM, 16 ADCA) | 12,533 | Gordon et al. [11] |
| AMLALL | 2 | 38 (27 ALL, 11 AML) | 7,129 | Golub et al. [12] |
| MLL | 3 | 57 (20 ALL, 17 MLL, 20 AML) | 12,582 | Armstrong et al. [13] |
| SRBCT | 4 | 63 (23 EWS, 8 BL, 12 NB, 20 RMS) | 2,308 | Khan et al. [14] |

expression of 2000 genes in 40 tumor biopsies ("negative") and 22 normal ("positive") biopsies from non-diseased colon biopsies from the same patients. An adult 2-class lung cancer set including 32 samples (16 malignant pleural mesothelioma (MPM) and 16 adenocarcinoma (ADCA)) of the lung with expression values for 12,533 genes [11] was also considered. Two leukemia data sets were evaluated: one 2-class data set with 38 arrays (27 ALL, 11 AML) containing expression for 7,129 genes [12], and the other consisting of 3 classes for 57 pediatric samples for lymphoblastic and myelogenous leukemia (20 ALL, 17 MLL and 20 AML) with expression values for 12,582 genes [13]. The Khan et al. [14] data set on pediatric small round blue-cell tumors (SRBCT) had expression profiles for 2,308 genes and 63 arrays comprising 4 classes (23 arrays for EWS-Ewing Sarcoma, 8 arrays for BL-Burkitt lymphoma, 12 arrays for NB-neuroblastoma, and 20 arrays for RMS-rhabdomyosarcoma). Values of gene expression within each microarray were log-transformed. Table 1 lists the datasets used allong with the number of classes and genes.

## 2.2  Gene Selection and Usage

Sets of $p = 32, 64, 128, 256$, and 512 genes were selected from the original high-dimensional datasets. Six methods were employed, and these are described in the following paragraphs.

**N(0,1)_BOOT_MUS.** This approach simulated sets of 32, 64, 128, 256, and 512 gene expression values with i.i.d. variates from the standard normal distribution, i.e., N(0,1). The random values of simulated gene expression were used for the in-bag arrays selected during bootstrapping (next section) prior to root MUSIC.

**RND_E_A_BOOT_MUS.** This method randomly selected sets of 32, 64, 128, 256, and 512 expression values ("E") from randomly selected arrays ("A") within the entire dataset. Randomly selected expression values were used for the in-bag arrays selected during bootstrapping prior to root MUSIC.

**RND_E_BOOT_MUS.** Sets of 32, 64, 128, 256, and 512 expression values ("E") were selected randomly within each array. Randomly selected expression values were used for the in-bag arrays selected during bootstrapping prior to root MUSIC.

**RND_F_BOOT_MUS.** Using this technique, sets of 32, 64, 128, 256, and 512 features ("F") or genes were randomly selected from the entire dataset, and used across all the arrays that were employed. Randomly selected genes were used for the in-bag arrays selected during bootstrapping prior to root MUSIC.

**RND_F.** This approach involve randomly sampling sets of 32, 64, 128, 256, and 512 features ("F") or genes from the full set of genes in each dataset, and then inputting them directly into the classification analysis. No bootstrapping was used and root MUSIC was not performed.

**BRN_BOOT_MUS.** For each dataset, all possible pairs of classes were used to identify for each 2-class comparison the genes with the greatest Gini index for 2-class discrimination. After obtaining a lists of ranked genes for all possible 2-class comparisons, a single non-redundant list of ranked genes was constructed. Sets of 32, 64, 128, 256, and 512 of the best ranked N ("BRN") genes were used for in-bag arrays selected during bootstrapping prior to root MUSIC.

**BRN.** Sets of 32, 64, 128, 256, and 512 of the best ranked genes ("BRN") for the full set of arrays (without bootstrapping) were used for classification analysis. The assumption surrounding use of the best ranked genes for input to classification runs is that class prediction should increase with the number of genes used. It also warrants noting that the intent here was not to determine classification accuracy as a function of incrementing the gene count by one, but rather to obtain a general trend as the bulk size of the genes used increases.

## 2.3   Bootstrap Root MUSIC Classifier

**Bootstrapping.** Let $\mathbf{x}_i$ ($i = 1, 2, \ldots, n$) be a $p$-dimensional microarray, and $B$ ($b = 1, 2, \ldots, B$) be the number of bootstraps. Bootstrapping involves selection of the same number of $n$ arrays in a dataset, but sampling with replacement is used. The probability that a microarray is not selected during bootstrapping is $(1 - 1/n)^n = \exp(-1) = 0.368$, and we used $n'$ to denote the number of unselected arrays, which varied over each bootstrap. Therefore, during each $b$th bootstrap, there were $n$ "in-bag" selected arrays $\mathbf{x}_i^{IB}$ in the in-bag dataset $\mathcal{D}_{IB}^{(b)}$, and $n'$ "out-of-bag" unselected arrays $\mathbf{x}_i^{OOB}$ in the out-of-bag dataset $\mathcal{D}_{OOB}^{(b)}$. The in-bag dataset was further partitioned into class-specific datasets $\mathcal{D}_{IB\omega}^{(b)}$ for root MUSIC analysis (next section). A total of 10 bootstraps ($B = 10$) was used.

**Superresolution Root MUSIC Dimension Reduction.** During each $b$th bootstrap, superresolution (root) MUSIC was employed on the bootstrap- and class-specific $p \times p$ covariance matrices $\mathbf{C}_\omega$ for random or best ranked 32, 64, 128, 256, and 512 genes using arrays in $\mathcal{D}_{IB\omega}^{(b)}$. The empirical eigenvalue distribution (e.e.d.) of $\mathbf{C}_\omega$ was determined using singular value decomposition. Let $\mathbf{E}_\omega = |\mathbf{E}_{n\omega}\mathbf{E}_{s\omega}|$ represent the partitioned matrix of eigenvectors of $\mathbf{C}_\omega$ in the noise and signal subspaces, whose eigenvalues (eigenvectors) are in ascending order ($\lambda_{(1)} \leq \lambda_{(2)} \leq \cdots \leq \lambda_{(p)}$). The eigenvector noise subspace was identified by applying the Marčenko-Pastur (MP) law [15] to the e.e.d. of $\mathbf{C}_\omega$. The MP law states that for a random matrix $\mathbf{X}$ having i.i.d. columns $(n, p \to \infty, \gamma = p/n)$, the minimum and maximum eigenvalues of its white Wishart matrix, $W_p(n, \boldsymbol{\Sigma}) = \mathbf{X}\mathbf{X}^T$, almost surely converge to $\lambda^- = \sigma^2(1 - \sqrt{\gamma})^2$ and $\lambda^+ = \sigma^2(1 + \sqrt{\gamma})^2$, respectively. The e.e.d. for $W_p(n, \boldsymbol{\Sigma})$ based on $\mathbf{X}$ with i.i.d. elements is given by

$$f(\lambda) = \max\left(0, 1 - \frac{1}{\gamma}\right)\delta(\lambda) + \frac{\sqrt{(\lambda^+ - \lambda)(\lambda - \lambda^-)}}{2\pi\gamma\lambda\sigma^2}I(\lambda^- \leq \lambda \leq \lambda^+), \quad (1)$$

where $\max()\delta(\lambda)$ represents the density at $\lambda = 0$ for the $p - n = p(1 - 1/\gamma)$ zero eigenvalues when $p > n$, i.e., $\gamma > 1$, and $I()$ represents the density when $\lambda$ is between $\lambda^-$ and $\lambda^+$. While $\lambda^+$ can be used as an estimate of the upper bound of the e.e.d of $W_p(n, \boldsymbol{\Sigma})$, the value of $\hat{\lambda}^+$ for $\mathbf{C}_\omega$ was obtained by fitting $f(\lambda)$ to the e.e.d. of $\mathbf{C}_\omega$ using particle swarm optimization [16]. Eigenvectors whose associated $\lambda$ was below $\hat{\lambda}^+$ were considered to be in $\mathbf{E}_{n\omega}$, while eigenvectors with $\lambda$ values above $\hat{\lambda}^+$ were assumed to be in $\mathbf{E}_{s\omega}$. Once the noise eigenvectors were determined, the predicted class-specific mode vector for an out-of-bag microarray $\mathbf{x}_i^{OOB}$ (during the $b$th bootstrap) was determined in the form

$$P(\mathbf{x}_i^{OOB}, \omega)^{(b)} = \frac{1}{(\mathbf{x}_i^{OOB})^T\mathbf{E}_{n\omega}\mathbf{E}_{n\omega}^T\mathbf{x}_i^{OOB}}. \quad (2)$$

Altogether, each mode vector for a microarray consisted of $\Omega$ values of $P(\mathbf{x}_i^{OOB}, \omega)^{(b)}$, which resulted in the row vector $\mathbf{P}_i^{(b)}$. For each OOB dataset containing $n'$ OOB arrays (during the $b$th bootstrap), the final matrix was

$$\mathbf{P}^{(b)}_{(n' \times \Omega)} = \begin{bmatrix} \mathbf{P}_1^{(b)} \\ \mathbf{P}_2^{(b)} \\ \vdots \\ \mathbf{P}_i^{(b)} \\ \vdots \\ \mathbf{P}_{n'}^{(b)} \end{bmatrix}, \quad (3)$$

which, when expanded to reveal individual elements appears as

$$\mathbf{P}^{(b)}_{(n' \times \Omega)} = \begin{bmatrix} P(\mathbf{x}_1, 1)^{(b)} & P(\mathbf{x}_1, 2)^{(b)} & \cdots & P(\mathbf{x}_1, \omega)^{(b)} & \cdots & P(\mathbf{x}_1, \Omega)^{(b)} \\ P(\mathbf{x}_2, 1)^{(b)} & P(\mathbf{x}_2, 2)^{(b)} & \cdots & P(\mathbf{x}_2, \omega)^{(b)} & \cdots & P(\mathbf{x}_2, \Omega)^{(b)} \\ \vdots & & & & & \\ P(\mathbf{x}_{n'}, 1)^{(b)} & P(\mathbf{x}_{n'}, 2)^{(b)} & \cdots & P(\mathbf{x}_{n'}, \omega)^{(b)} & \cdots & P(\mathbf{x}_{n'}, \Omega)^{(b)} \end{bmatrix}. \quad (4)$$

The resulting $\mathbf{P}^{(b)}$ matrices from the $B = 10$ bootstraps were stacked on top of one another using the following form

$$
\underset{(N' \times \Omega)}{\mathbf{P}} = \begin{bmatrix} \mathbf{P}^{(1)} \\ \mathbf{P}^{(2)} \\ \vdots \\ \mathbf{P}^{(b)} \\ \vdots \\ \mathbf{P}^{(B)} \end{bmatrix}, \tag{5}
$$

which was input into classification analysis. Note that the number of rows $N' = n'^{(1)} + n'^{(2)} + \cdots + n'^{(B)}$ in $\mathbf{P}$ was equal to the sum of the varying number of OOB arrays obtained from each bootstrap. Algorithm 1 lists the workflow for Bootstrap Root Music (BRM) described above.

## 2.4   Classification Runs Using Bootstrap Root MUSIC (BRM)

For each dataset, classification analysis was performed on either (a) the set of predicted mode vectors $\mathbf{P}$ for OOB arrays, or (b) the original $n$ arrays using sets of 32, 64, 128, 256, or 512 best ranked N genes without bootstrapping and without root MUSIC. Classification [17] runs were made using k-nearest neighbor (KNN), learning vector quantization (LVQ1), and kernel regression (KREG). Ten 10-fold cross validation (CV) was used to establish classification accuracy, or performance [2]. During each of the ten repartitionings, arrays were randomly shuffled and assigned to the 10 folds prior to classification analysis. The confusion matrix was constructed by assigning arrays into rows for the predicted class and columns for the true class label, and accuracy was determined as the ratio of the trace of the confusion matrix to the sum of all elements after the ten repartitions. In summary, when bootstrapping and root MUSIC was employed, the input sets of 32, 64, 128, 256, 512 genes for in-bag arrays were used to extract class-specific noise eigenvalues, which were applied to OOB arrays in (2) to predict $\Omega$-dimensional mode vectors in $\mathbf{P}$. The $\mathbf{P}$ for all $N'$ OOB arrays over $B = 10$ bootstraps were then used for input during classification analysis. Otherwise, full sets of 32, 64, 128, 256, or 512 genes for all $n$ arrays in a dataset were input into classification analysis.

## 3   Results

The overall average classification accuracy for the various feature selection methods is listed in Table 2. As can be seen, bootstrap root MUSIC (BRM) of randomly selected within-array expression (RND_E_BOOT_MUS) increased, on average, classification accuracy when compared with results based on random standard normal variates (N(0,1)_BOOT_MUS) or random selection of expression from any array (RND_E_A_BOOT_MUS). On average, the use of BRM

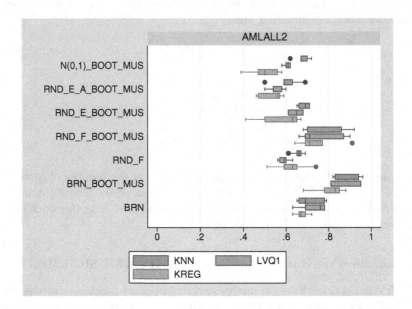

**Fig. 1.** Ten 10-fold CV classification accuracy for 2-class AMLALL dataset.

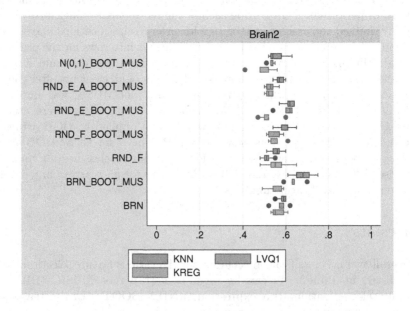

**Fig. 2.** Ten 10-fold CV classification accuracy for 2-class Brain dataset.

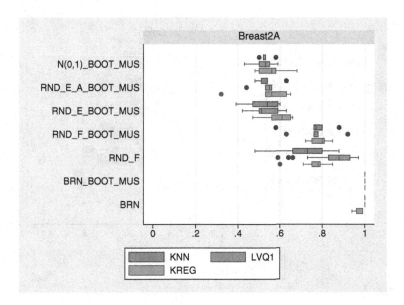

**Fig. 3.** Ten 10-fold CV classification accuracy for 2-class BreastA dataset.

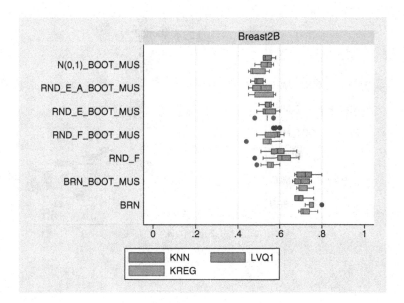

**Fig. 4.** Ten 10-fold CV classification accuracy for 2-class BreastB dataset.

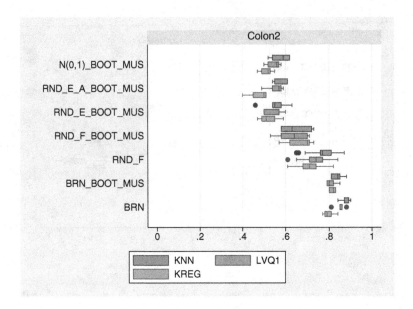

**Fig. 5.** Ten 10-fold CV classification accuracy for 2-class Colon dataset.

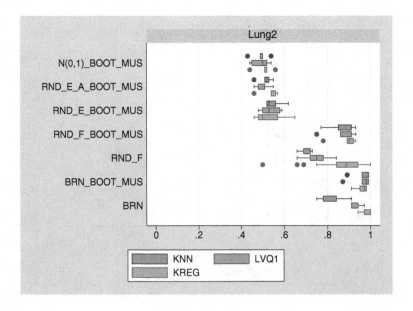

**Fig. 6.** Ten 10-fold CV classification accuracy for 2-class Lung dataset.

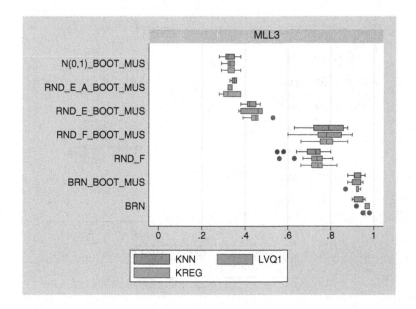

**Fig. 7.** Ten 10-fold CV classification accuracy for 3-class MLL dataset.

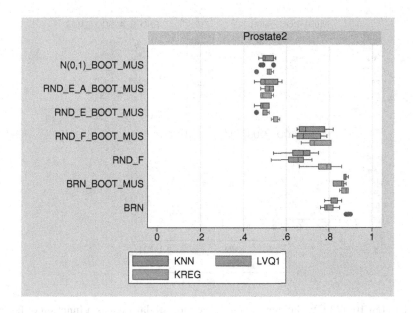

**Fig. 8.** Ten 10-fold CV classification accuracy for 2-class Prostate dataset.

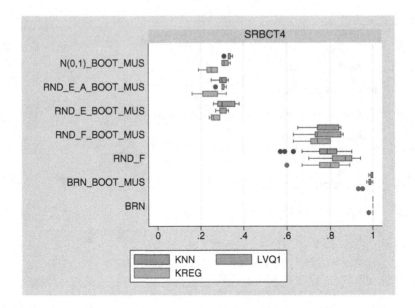

**Fig. 9.** Ten 10-fold CV classification accuracy for 4-class SRBCT dataset.

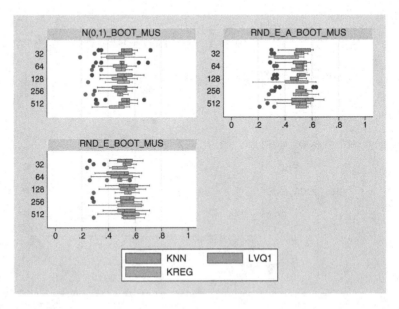

**Fig. 10.** Ten 10-fold CV classification accuracy for all datasets as a function of feature selection and number of genes.

---

### Algorithm 1. BRM: Bootstrap Root MUSIC

**Data**: $n$ arrays, $p$=32,64,128,256,512 genes; arrays $\mathbf{x}_i$ ($i = 1, 2, \ldots, n$); $\Omega$ classes

**for** $b \leftarrow 1$ **to** #*Bootstraps* **do**

    Randomly select $p$ genes (or use $p$ best ranked genes)

    Generate bootstrap dataset $\mathcal{D}_{IB}^{(b)}$ containing $n$ "in-bag" arrays $\mathbf{x}_i^{IB}$

    Separate in-bag arrays into class-specific datasets $\mathcal{D}_{IB\omega}^{(b)}$

    Assign the $n'$ unselected "out-of-bag" arrays $\mathbf{x}_i^{OOB}$ to dataset $\mathcal{D}_{OOB}^{(b)}$

    **for** $\omega \leftarrow 1$ **to** $\Omega$ **do**

        Generate $p \times p$ covariance matrix $\mathbf{C}_\omega$ from $\mathcal{D}_{IB\omega}^{(b)}$

        Perform root MUSIC on $\mathbf{C}_\omega$ to yield $\mathbf{E}_{n\omega}$

        **for** $i \leftarrow 1$ **to** $n'$ **do**

            Calculate mode-vector $P(\mathbf{x}_i^{OOB}, \omega)^{(b)}$ from $\mathbf{E}_{n\omega}$ using (2)

Perform ten 10-fold CV classification using $\mathbf{P}$ as input

---

resulted in greater accuracy results when comparing use of random features (i.e., RND_F_BOOT_MUS vs. RND_F) and best ranked N genes (BRN_BOOT_MUS vs. BRN). For the KNN classifier, use of the BRM approach based on random features (i.e., RND_F_BOOT_MUS vs. RND_F) resulted in an accuracy that was 3.5 % greater than random gene selection alone (t-test, $P = 0.026$). Tables 3–5 list the classification accuracies obtained for the KNN, LVQ1, and KREG classifiers, respectively, for each dataset. Tail probabilities for t-tests for the equality of mean accuracy from BRM based on random gene selection (RND_F_BOOT_MUS) and random genes input directly into the classifier (RND_F) are listed in the far right column of Tables 3–5.

Figures 1–9 show for each dataset the ten 10-fold CV accuracy for the three classifiers employed. It was observed, that whenever classification accuracy for using the best rank N features directly input into classifiers ("BRN") was greater than accuracy based on BRM of best ranked N genes ("BRN_BOOT_MUS"), classification accuracy for RND_F runs exceeded accuracy for RND_F_BOOT_MUS

**Table 2.** Mean ± s.d. of classification accuracy for classifiers used as a function of feature selection method.

| Feature selection | KNN | LVQ1 | KREG | n |
|---|---|---|---|---|
| N(0,1)_BOOT_MUS | 0.506 ± 0.112 | 0.486± 0.099 | 0.462± 0.107 | 45 |
| RND_E_A_BOOT_MUS | 0.496 ± 0.107 | 0.480± 0.096 | 0.466± 0.114 | 45 |
| RND_E_BOOT_MUS | 0.522± 0.110 | 0.515± 0.105 | 0.504± 0.108 | 45 |
| RND_F_BOOT_MUS | 0.720*± 0.114 | 0.710± 0.127 | 0.714± 0.123 | 45 |
| RND_F | 0.685*± 0.088 | 0.700± 0.124 | 0.712± 0.116 | 450 |
| BRN_BOOT_MUS | 0.878± 0.114 | 0.862± 0.124 | 0.841± 0.138 | 45 |
| BRN | 0.828± 0.140 | 0.845± 0.141 | 0.840± 0.151 | 45 |

*T-test for equality of means of KNN accuracy 0.72 (RND_F_BOOT_MUS) vs. 0.685 (RND_F), significant with $P = 0.026$. T-test assumed unequal variances.

**Table 3.** Mean ± s.d. of KNN classification accuracy for classifiers used as a function of datasets for randomly selected genes with and without use of BRM.

| Dataset | RND_F_BOOT_MUS $\mu \pm \sigma$ | RND_F $\mu \pm \sigma$ | P-value |
|---|---|---|---|
| AMLALL2 | 0.772±0.110 | 0.659±0.018 | <0.0005 |
| Brain2 | 0.598±0.041 | 0.559±0.022 | 0.00108 |
| Breast2A | 0.758±0.110 | 0.725±0.093 | 0.46389 |
| Breast2B | 0.582±0.011 | 0.590±0.043 | 0.68813 |
| Colon2 | 0.648±0.073 | 0.771±0.048 | <0.0005 |
| Lung2 | 0.862±0.063 | 0.693±0.023 | <0.0005 |
| MLL3 | 0.776±0.103 | 0.720±0.052 | 0.04343 |
| Prostate2 | 0.720±0.076 | 0.669±0.052 | 0.04902 |
| SRBCT4 | 0.764±0.083 | 0.779±0.075 | 0.67806 |

RND_F_BOOT_MUS - Uses bootstrap root MUSIC (BRM) on randomly selected features prior to classification.
RND_F - Directly inputs randomly selected features into classification.

**Table 4.** Mean ± s.d. of LVQ1 classification accuracy for classifiers used as a function of datasets for randomly selected genes with and without use of BRM.

| Dataset | RND_F_BOOT_MUS $\mu \pm \sigma$ | RND_F $\mu \pm \sigma$ | P-value |
|---|---|---|---|
| AMLALL2 | 0.766±0.111 | 0.590±0.020 | <0.0005 |
| Brain2 | 0.552±0.035 | 0.514±0.016 | 0.00005 |
| Breast2A | 0.774±0.103 | 0.864±0.087 | 0.03407 |
| Breast2B | 0.566±0.054 | 0.738±0.050 | 0.00007 |
| Colon2 | 0.632±0.077 | 0.738±0.050 | 0.00007 |
| Lung2 | 0.862±0.070 | 0.753±0.042 | <0.0005 |
| MLL3 | 0.774±0.115 | 0.731±0.047 | 0.09744 |
| Prostate2 | 0.702±0.070 | 0.645±0.051 | 0.02325 |
| SRBCT4 | 0.762±0.095 | 0.857±0.059 | 0.00215 |

RND_F_BOOT_MUS - Uses bootstrap root MUSIC (BRM) on randomly selected features prior to classification.
RND_F - Directly inputs randomly selected features into classification.

runs. Another observation was that, when best ranked N genes were used (Figs. 10 and 11), the interquartile range of classification accuracy was smaller when bootstrapping and MUSIC was used when compared with direct input of best ranked genes into classifiers.

**Table 5.** Mean ± s.d. of KREG classification accuracy for classifiers used as a function of datasets for randomly selected genes with and without use of BRM.

| Dataset | RND_F_BOOT_MUS $\mu \pm \sigma$ | RND_F $\mu \pm \sigma$ | P-value |
|---|---|---|---|
| AMLALL2 | 0.744±0.104 | 0.627±0.052 | 0.00006 |
| Brain2 | 0.556±0.035 | 0.559±0.036 | 0.85954 |
| Breast2A | 0.786±0.051 | 0.773±0.041 | 0.50806 |
| Breast2B | 0.536±0.063 | 0.555±0.025 | 0.17570 |
| Colon2 | 0.666±0.068 | 0.716±0.045 | 0.0279 |
| Lung2 | 0.884±0.060 | 0.871±0.097 | 0.77181 |
| MLL3 | 0.776±0.081 | 0.736±0.038 | 0.04996 |
| Prostate2 | 0.746±0.062 | 0.778±0.046 | 0.15608 |
| SRBCT4 | 0.736±0.071 | 0.789±0.071 | 0.11359 |

RND_F_BOOT_MUS - Uses bootstrap root MUSIC (BRM) on randomly selected features prior to classification.
RND_F - Directly inputs randomly selected features into classification.

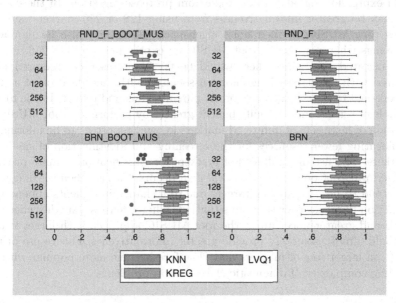

**Fig. 11.** Ten 10-fold CV classification accuracy for all datasets as a function of feature selection and number of genes.

## 4    Discussion

We have shown that BRM exploits the noise eigenvectors of the class-specific gene-by-gene covariance matrix of gene expression and when performed on randomly selected genes can improve results. BRM when performed on the best ranked genes also resulted in lower variance across data sets and less bias depending on the filtration method used. It is not surprising that class-specific root MUSIC is capable of increased information retrieval from noise eigenvectors, as the noise for a given class is unique, and there are typically many more eigenvectors in the noise region than there are in the signal region. BRM can also reduce uncertainty of classifier accuracy based on randomly selected genes and best ranked genes. In another sense, use of the noise region of eigenspace for classification basically exploits what is not known about a given class of arrays. Since there are many more noise eigenvectors, there is essentially more information provided when compared with the signal when exploited by root MUSIC.

The majority of groups now trying to identify reduced sets of markers for development of diagnostic chips are focusing on parsimony and robustness. Parsimony attempts to identify fewer genes yielding greater classification performance in order to optimize resource allocation. The parallel requirement for robustness results in the identification of genes that perform well across different environments. Unfortunately, parsimony and robustness are not always attainable for certain diagnostic classes or certain biomaterials such as urine and saliva, where protein expression signals can degrade from protease digestion. In these cases, there may be greater informativeness from using 128 or 256 markers and performing root MUSIC on their noise eigenspace. There may also be significant cost savings when performing root MUSIC on expression of 128 or 256 markers on a fabricated chip, since there would be far fewer resources required when compared with the painstaking efforts associated with identification of a small set of optimal markers which are jointly parsimonious and robust. In the future, the whole of science may actually benefit greater by performing root MUSIC on expression of large sets of unfiltered markers known (reported) to be informative and allocating fewer resources for the screening and identification of small sets of markers. The use of small sets of markers for classification is more prone to lose discriminatory power when *concept drift* occurs, that is, when the diagnostic classes tend to become moving targets due to a change in patients appended to a prediction database over time. In situations where there is a smaller effect size (signal) and greater uncertainty, the root MUSIC approach with larger marker sets would almost assuredly provide greater informativenes. Future use of root MUSIC on larger sets of markers may therefore become more popular with the increasing complexity of diagnostic classification problems.

# References

1. Ambroise, C., McLachlan, G.J.: Selection bias in gene extraction on the basis of microarray gene-expression data. PNAS **99**(10), 6562–6566 (2002)
2. Kohavi, R.: A study of cross-validation and bootstrap for accuracy estimation and model selection. In: Proceedings of the International Joint Conference on Artificial Intelligence (IJCAI), pp. 1137–1145 (1995)
3. Reddi, S.S.: Multiple source location-a digital approach. IEEE Trans. Aerosp. Electron. Syst. **AES–15**(1), 95–105 (1979)

4. Schmidt, R.O.: Multiple emitter location and signal parameter estimation. IEEE Trans. Antennas Propag. **AP–34**(3), 276–280 (1986)
5. Kim, K.-T., Seo, D.K., Kim, H.-T.: Efficient radar target recognition using the MUSIC algorithm and invariant features. IEEE Trans. Antennas Propag. **50**(3), 325–337 (2002)
6. Pomeroy, S.L., Tamayo, P., Gaasenbeek, M., Sturla, L.M., Angelo, M., McLaughlin, M.E., Kim, J.-Y.H., Goumnerovak, L.C., Blackk, P.M., Lau, C., Allen, J.C., ZagzagI, D., Olson, J.M., Curran, T., Wetmore, C., Biegel, J.A., Poggio, T., Mukherjee, S., Rifkin, R., Califanokk, A., Stolovitzkykk, G., Louis, D.N., Mesirov, J.P., Lander, E.S., Golub, T.R.: Prediction of central nervous system embryonal tumour outcome based on gene expression. Nature **415**(6870), 436–442 (2002)
7. Singh, D., Febbo, P.G., Ross, K., Jackson, D.G., Manola, J., Ladd, C., Tamayo, P., Renshaw, A.A., D'Amico, A.V., Richie, J.P., Lander, E.S., Loda, M., Kantoff, P.W., Golub, T.R., Sellers, W.R.: Gene expression correlates of clinical prostate cancer behavior. Cancer Cell. **1**(2), 203–209 (2002)
8. Hedenfalk, I., Duggan, D., Chen, Y., et al.: Gene-expression profiles in hereditary breast cancer. N. Engl. J. Med. **344**, 539–548 (2001)
9. van 't Veer, L.J., Dai, H., van de Vijver, M.J., He, Y.D., Hart, A.A., Mao, M., Peterse, H.L., van der Kooy, K., Marton, M.J., Witteveen, A.T., Schreiber, G.J., Kerkhoven, R.M., Roberts, C., Linsley, P.S., Bernards, R., Friend, S.H.: Gene expression profiling predicts clinical outcome of breast cancer. Nature **415**, 530–536 (2002)
10. Alon, U., Barkai, N., Notterman, D.A., Gish, K., Ybarra, S., Mack, D., Levine, A.J.: Broad patterns of gene expression revealed by clustering of tumor and normal colon tissues probed by oligonucleotide arrays. Proc. Natl. Acad. Sci. USA **96**(12), 6745–6750 (1999)
11. Gordon, G.J., Jensen, R.V., Hsiao, L.L., Gullans, S.R., Blumenstock, J.E., Ramaswamy, S., Richards, W.G., Sugarbaker, D.J., Bueno, R.: Translation of microarray data into clinically relevant cancer diagnostic tests using gene expression ratios in lung cancer and mesothelioma. Cancer Res. **62**(17), 4963–5967 (2002)
12. Golub, T.R., Slonim, D.K., Tamayo, P., Huard, C., Gaasenbeek, M., Mesirov, J.P., Coller, H., Loh, M., Downing, J.R., Caligiuri, M.A., Bloomfield, C.D., Lander, E.S.: Molecular classification of cancer: class discovery and class prediction by gene expression. Science **286**, 531–537 (1999)
13. Armstrong, S.A., Staunton, J.E., Silverman, L.B., Pieters, R., den Boer, M.L., Minden, M.D., Sallan, S.E., Lander, E.S., Golub, T.R., Korsmeyer, S.J.: MLL translocations specify a distinct gene expression profile that distinguishes a unique leukemia. Nat. Genet. **30**(1), 41–47 (2001)
14. Khan, J., Wei, J.S., Ringner, M., Saal, L.H., Ladanyi, M., Westermann, F., Berthold, F., Schwab, M., Antonescu, C.R., C. Peterson, C.R., Meltzer, R.S.: Classification and diagnostic prediction of cancers using gene expression profiling and artificial neural networks. Nat. Med. **7**, 673–679 (2001)
15. Marčenko, V.A., Pastur, L.A.: Mat. Sb., (N.S.) 72(114), 507–536 (1967)
16. Kennedy, J., Eberhart, R.C., Particle swarm optimization. In: Proceedings of IEEE International Conference on Neural Networks, pp. 1942–1948. IEEE Press, Piscataway (1995)
17. Peterson, L.E.: Classification Analysis of DNA Microarrays. John Wiley and Sons, New York (2013)

# Special Session: Data Integration and Analysis in Omic-Science

# Prediction of Single-Nucleotide Polymorphisms Causative of Rare Diseases

Maria Brigida Ferraro[1]([✉]) and Mario Rosario Guarracino[2]

[1] Department of Statistical Sciences, Sapienza University of Rome, Rome, Italy
mariabrigida.ferraro@uniroma1.it
[2] High Performance Computing and Networking Institute,
National Research Council, Naples, Italy

**Abstract.** The study of rare diseases uses next-generation sequencing (NGS) technology to detect causative mutations in the human genome. NGS is a new approach for biomedical research, useful for the genetic diagnosis in extremely heterogeneous conditions. Nevertheless, only few publications address the problem when pooled experiments are considered, and existing tools are often inaccurate. In this work we focus on rare diseases and we describe how data are generated by NGS.

We present how data are organized in the pre-processing phase, how they are filtered and features constructed in the learning phase. We compare different computational procedures to identify and classify variants potentially related to rare diseases and we biologically validate the obtained results.

## 1 Introduction

The introduction of capillary electrophoresis (CE)-based Sanger sequencing makes possible the interpretation of genetic information from any given biological system. Although this technology has been widely used, it has some limitations with respect to throughput, scalability, speed, and resolution. In order to overcome these drawbacks, Next-Generation Sequencing (NGS) has been introduced [1,2]. In principle, the idea of NGS technology is similar to CE: the bases of a small fragment of DNA are sequentially identified from signals emitted as each fragment is re-synthesized from a DNA template strand. NGS extends this process across millions of reactions through a massive parallelization, rather than being limited to a single or a few DNA fragments. It provides an enormous number of reads, which permits the sequencing of entire genomes at a fraction of the costs for Sanger technology. Hence, this allows to get the complete genomic sequences for a large number of individuals. We are interested in molecular diagnosis of muscular diseases: neuromuscular disorders and progressive loss of motor function [3,4]. This problem is characterized by genetic heterogeneity (see, for more details, [5]). In this study 98 genes related to these diseases are considered.

In heterogeneous genetic conditions, about 40 % of patients do not obtain a molecular diagnosis by means of a traditional approach, because it would require a high number of gene sequences, which is expensing and time-consuming.

© Springer International Publishing Switzerland 2014
E. Formenti et al. (Eds.): CIBB 2013, LNBI 8452, pp. 213–224, 2014.
DOI: 10.1007/978-3-319-09042-9_15

On the other hand, NGS is useful for the molecular diagnosis of extremely hetero-geneous conditions. For genetic diagnosis, it provides an easy analytic pipeline, high specificity and sensitivity. Furthermore, it reduces the overall execution time.

Currently NGS techniques are mainly used to sequence individual genomes and the costs for population-scale analyses are still too high. To this extent, pooling individuals in NGS experiments can reduce the costs, still maintaining an adequate coverage to detect single nucleotide polymorphism (SNP) in each patient. Furthermore, pooled NGS is often theoretically more effective in muta-tion discovery and provides more accurate allele frequency estimates [6]. In prac-tice, only few publications address this problem and existing tools are often inaccurate.

Calvo *et al.* [7] introduce high-throughput, pooled sequencing to identify mutations in NUBPL and FOXRED1 in human complex I deficiency. This prob-lem is characterized by a large number of both mitochondrial (mt) and nuclear genes. Seven pools of DNA from a group of 103 cases and 42 healthy controls are involved. The aim is to identify 151 rare variants that are predicted to affect protein function. In this case, genetic diagnoses are established in 13 of 60 previ-ously unsolved cases. This study illustrates how large-scale sequencing, coupled with functional prediction and experimental validation, can be used to identify causal mutations in individual cases. Unfortunately, they were able to confirm only the mutations only in half of the 103 subjects.

Wang *et al.* [8] propose sequencing pooled mtDNA of multiple individuals for estimating allele frequency using the Illumina genome analyzer (GA) II sequenc-ing system. Each pool includes mtDNA samples of 20 subjects that have been previously sequenced using Sanger sequencing. Furthermore, each pool is repli-cated to assess variation of the sequencing error between pools. The proposed technique is not resilient to sequencing errors, thus providing a large number of false positives.

Finally, Ding *et al.* [9] compare four standard supervised machine learning algorithms to predict causative SNP in tumour/normal NGS experiments. In order to evaluate these approaches (random forest, Bayesian additive regression tree, support vector machine and logistic regression), features are constructed to represent 3369 candidate somatic SNPs from 48 breast cancer genomes, originally predicted with naive methods and subsequently revalidated to establish ground truth labels. The solution depends on third-party software packages.

The aim of this paper is to predict causative mutations for rare diseases in pooled experiments. The genomic regions are analyzed by means of the Agi-lent HaloPlex Target Enrichment System [10]. In a preliminary phase the data are filtered by means of classification rules. Then, features are constructed and subsequently used in the learning phase. The last part of the computational procedure consists of a supervised classification of potential mutations.

The paper is organized as follows. Next section deals with formats and tools for managing NGS data, pooling strategy and its empirical analysis. In Sect. 3 a computational procedure to identify and predict causative mutations for rare diseases is introduced and discussed. In Sect. 4 results are discussed. Finally, in Sect. 5 some concluding remarks and open problems are addressed.

## 2 Materials

### 2.1 Formats and Tools for Managing NGS Data

The advent of high-throughput DNA sequencing in biological sciences has enabled researchers to obtain millions of short sequence reads in a single, low-cost experiment. Instead of one whole genome, that is characterized by billions of nucleotides, these sequence reads are, in general, very short, usually 36–200 nucleotides. In addition, they are very redundant, namely, they may have the same sequence. The sequences can be used for detecting small differences in the genome of the sample. This is possible by means of computational techniques that align each short sequence read against a reference genome. As second step, the obtained alignments are analyzed and, finally, the positions where these differences emerge from are determined. Many software have been created for alignment. Most NGS alignment programs produce a standard output file in Sequence Alignment/Map format (SAM) [11]. The mandatory fields in the SAM format and their descriptions are reported in Table 1. In details, we want to focus on MAPQ and QUAL. The first one indicates the mapping quality and it is the phred-scaled error probability which is equal to $-10\log_{10} P$, where $P$ is the error probability in mapping. This is also known as *alignment quality score*. The second one is the ASCII encoding of base call QUALity plus 33. A base call quality is equal to $-10\log_{10} P$, where $P$ is the probability of an incorrect base call. Since the SAM files are large, they can be converted to the binary equivalent BAM files, making genetic analyses less storage consuming.

### 2.2 Pooling

In this study, 8 pools of 16 patients each one are considered.

**Table 1.** Mandatory fields in the SAM format

| No. | Name | Description |
|---|---|---|
| 1 | QNAME | Query NAME of the read or the read pair |
| 2 | FLAG | Bitwise FLAG (pairing, strand, mate strand, etc.) |
| 3 | RNAME | Reference sequence NAME |
| 4 | POS | 1-Based leftmost POSition of clipped alignment |
| 5 | MAPQ | MAPping Quality (Phred-scaled) |
| 6 | CIGAR | Extended CIGAR string (operations: MIDNSHP) |
| 7 | MRNM | Mate Reference NaMe (= if same as RNAME) |
| 8 | MPOS | 1-Based leftmost Mate POSition |
| 9 | ISIZE | Inferred Insert SIZE |
| 10 | SEQ | Query SEQuence on the same strand as the reference |
| 11 | QUAL | Query QUALity (ASCII-33=Phred base quality) |

**Table 2.** Pooling: the original pools (1–8) and the replicated ones (9–16). The control patients are indicated in bold.

| Pool | | | | | | | | | | | | | | | | |
|---|---|---|---|---|---|---|---|---|---|---|---|---|---|---|---|---|
| #1 | **1** | 2 | 3 | 4 | 5 | 6 | 7 | 8 | 9 | 10 | 11 | 12 | 13 | 14 | 15 | **16** |
| #2 | 17 | 18 | **19** | 20 | 21 | 22 | 23 | 24 | 25 | 26 | 27 | 28 | 29 | **30** | 31 | 32 |
| #3 | 33 | 34 | 35 | 36 | **37** | 38 | 39 | 40 | 41 | 42 | 43 | **44** | 45 | 46 | 47 | 48 |
| #4 | 49 | 50 | 51 | 52 | 53 | 54 | **55** | 56 | 57 | **58** | 59 | 60 | 61 | 62 | 63 | 64 |
| #5 | 65 | 66 | 67 | 68 | 69 | 70 | 71 | **72** | **73** | 74 | 75 | 76 | 77 | 78 | 79 | 80 |
| #6 | 81 | 82 | 83 | 84 | 85 | **86** | 87 | 88 | 89 | 90 | **91** | 92 | 93 | 94 | 95 | 96 |
| #7 | 97 | 98 | 99 | **100** | 101 | 102 | 103 | 104 | 105 | 106 | 107 | 108 | **109** | 110 | 111 | 112 |
| #8 | 113 | **114** | 115 | 116 | 117 | 118 | 119 | 120 | 121 | 122 | 123 | 124 | 125 | 126 | **127** | 128 |
| #9 | 1 | 2 | 17 | 18 | 33 | 34 | 49 | 50 | 65 | 66 | 81 | 82 | 97 | 98 | 113 | **114** |
| #10 | 3 | 4 | **19** | 20 | 35 | 36 | 51 | 52 | 67 | 68 | 83 | 84 | 99 | **100** | 115 | 116 |
| #11 | 5 | 6 | 21 | 22 | **37** | 38 | 53 | 54 | 69 | 70 | 85 | **86** | 101 | 102 | 117 | 118 |
| #12 | 7 | 8 | 23 | 24 | 39 | 40 | **55** | 56 | 71 | **72** | 87 | 88 | 103 | 104 | 119 | 120 |
| #13 | 9 | 10 | 25 | 26 | 41 | 42 | 57 | **58** | **73** | 74 | 89 | 90 | 105 | 106 | 121 | 122 |
| #14 | 11 | 12 | 27 | 28 | 43 | **44** | 59 | 60 | 75 | 76 | **91** | 92 | 107 | 108 | 123 | 124 |
| #15 | 13 | 14 | 29 | **30** | 45 | 46 | 61 | 62 | 77 | 78 | 93 | 94 | **109** | 110 | 125 | 126 |
| #16 | 15 | **16** | 31 | 32 | 47 | 48 | 63 | 64 | 79 | 80 | 95 | 96 | 111 | 112 | **127** | 128 |

Each pool contains 2 control patients with known mutations. Pools are replicated, using 2 patients from each original pool. The pools are reported in Table 2. Each row represents a pool, where patients are numbered from 1 to 128.

A control heterozygous mutation is present in each of two samples of every pool. This means that it is present in 1 out 32 alleles, that is, approximately 3.12 % of reads. The total number of generated reads and their qualities are reported in Table 3. It is worth noting that the percentage of reads with a mapping quality greater than $Q30$, that is, a probability of mapping error lower than 0.001, is always greater than 90 %.

## 3    Methods

### 3.1    Decision Rules

The proposed computational procedure consists of two parts: a rules based filtering and a supervised classification algorithm. In the first step, for each position, the frequencies of four bases are computed together with the coverage, that is the sum of the frequencies. Since we want approximately an average of 30 reads for each individual, that is 480 reads, we select only the positions with a coverage at least equal to 500 (we use 500 to approximate 480). Furthermore, taking into account that the contribution of each sample varies from 1 to 2.5 and fixing an error rate equal to 0.01, a mutation is characterized by one base with a frequency in the interval (1.1 %, 8 %) and two bases below 1 %. The obtained dataset is classified by means of supervised techniques.

**Table 3.** Total number of reads, percentage of reads with a mapping quality $\geq Q30$ and mean quality score ($QS$) for each pool.

| ID pool | Number of reads (millions) | $\% \geq Q30$ | Mean $QS$ |
|---------|---------------------------|---------------|-----------|
| #1      | 35.970                    | 90.47         | 35.7      |
| #2      | 32.522                    | 90.48         | 35.7      |
| #3      | 25.408                    | 90.01         | 35.52     |
| #4      | 27.523                    | 90.33         | 35.62     |
| #5      | 24.384                    | 90.23         | 35.59     |
| #6      | 21.221                    | 89.78         | 35.41     |
| #7      | 28.266                    | 89.80         | 35.44     |
| #8      | 38.356                    | 93.52         | 36.78     |
| #9      | 17.690                    | 93.59         | 36.85     |
| #10     | 9.995                     | 93.2          | 36.68     |
| #11     | 28.823                    | 93.51         | 36.82     |
| #12     | 17.229                    | 93.8          | 36.96     |
| #13     | 41.533                    | 92.92         | 36.45     |
| #14     | 48.657                    | 93.19         | 36.59     |
| #15     | 46.022                    | 93.24         | 36.62     |
| #16     | 27.419                    | 93.1          | 36.55     |

## 3.2   Features

The next step consists in constructing the features used in the learning phase of the classification procedure. The supervised classification is based on a training and a testing set. In the training phase, each classifier learns from the data on the basis of some features. In this study, taking into account that the aim is to predict causative mutations, we construct features mainly related to the base frequency, the mapping quality and the base quality. In details, for each position we consider: the frequency of each base ('T', 'G', 'C', 'A'), the minimum, the maximum and the average mapping quality, and the minimum, the maximum and the average base quality.

## 3.3   Generalized Eigenvalue Classification

Mangasarian and Wild [12] introduce a multisurface proximal support vector machine classification via generalized eigenvalues (GEPSVM). Each plane is obtained as the closest to one class, and as far as possible from the other one (see Fig. 1). The solution plane $x^T\omega - \gamma = 0$ for class $A$ is the one that is the closest to that class, and the furthest from $B$:

$$\min_{\omega, \gamma \neq 0} \frac{\|A\omega - e\gamma\|^2}{\|B\omega - e\gamma\|^2}, \tag{1}$$

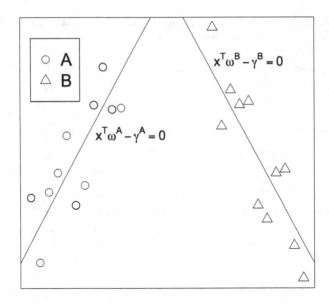

**Fig. 1.** Example of two classes, A and B, and the planes obtained by GEPSVM

where $e$ is a unit column vector of proper dimension

The optimization problem to obtain the planes is reduced to the minimization of a Rayleigh quotient, whose solution is obtained by solving a generalized eigenvalue problem, for which many well known results exist in literature [13,14].

Let $G = [A \quad - e]^T [A \quad - e]$ and $H = [B \quad - e]^T [B \quad - e]$, then the problem (1) becomes:

$$\min_{z \neq 0} \frac{z^T G z}{z^T H z}, \tag{2}$$

with $z = [\omega^T \quad \gamma]^T$. This is the Rayleigh quotient of the generalized eigenvalue problem $Gz = \lambda H z$. The inverse of the objective function in (2) has the same eigenvectors and reciprocal eigenvalues. In [12] it is proven that proximal planes are defined by

$$z_{min} = [\omega_1 \ \gamma_1]', \quad z_{max} = [\omega_{-1} \ \gamma_{-1}]' \tag{3}$$

where $z_{min}$ and $z_{max}$ are the eigenvectors related to the eigenvalues of smallest and largest modulo, respectively.

The model is also able to predict the class label of unknown samples by assigning them to the class with minimum distance from the hyperplane. For a point $x_u$:

$$class(x_u) = \arg \min_{i \in \{A,B\}} = \frac{|w_i^T x_u - \gamma_i|}{\|w_i\|}.$$

The matrices involved in the objective function can be singular, hence the solution can be not unique. Mangasarian and Wild [12] propose a Tikhonov regularization term in two generalized eigenvalue problems. Guarracino *et al.* [15]

introduce a new regularization technique (ReGEC) in order to solve only one eigenvalues problem. In Cifarelli *et al.* [16] the incremental version of the algorithm is proposed.

We use ReGEC for the prediction of SNPs. We use the set of true mutations and non-mutation positions as a training set in the learning phase and then we test the positions obtained in the filtering step in order to predict mutations.

# 4    Results and Discussion

## 4.1    Benchmark Dataset

We constructed a benchmark dataset composed of 16 files from lines of a real experiments. The benchmark is based on 148 mutations and 57 non mutations biologically verified. The files are built using all reads of the original experiments that have been mapped starting at most 75 bases before the 148 + 57 testing positions.

## 4.2    Other Tools Results

In this section we briefly recall four tools existing in literature used for NGS data: the Genome Analysis Toolkit (GATK), SNVer, FreeBayes and CRISP. These are not directly comparable with supervised classification techniques but they are very used in this context. GATK is a software package developed at the Broad Institute to analyse next-generation data [17,18]. Among the different tools offered, the most important is variant discovery and genotyping. The GATK was designed using the functional programming paradigm of MapReduce. It is characterized by a robust architecture.

SNVer is a statistical tool for calling common and rare variants in analysis of pool or individual next-generation sequencing data [19]. Loci with any (low) coverage can be tested and depth of coverage will be quantitatively factored into final significance calculation.

FreeBayes is a Bayesian genetic variant detector designed to find small polymorphisms, specifically SNPs, Indels, and MNPs (multi-nucleotide polymorphisms) smaller than the length of a short-read [20].

CRISP uses a cross-pool comparison approach to distinguish sequencing errors from rare variants. It can be used to evaluate pooled sequencing datasets (human and bacterial) generated by the Illumina sequencing platform [21].

We check the positions of the benchmark dataset of mutations by means of all above software. GATK recognizes 104 out of 148 as true positive and 44 as false negative. Only 4 out of 57 are classified as false positive. The number of false negative obtained with FreeBayes is 82 out of 148, while there are no false positives. SNVer has 0 % specificity, and CRISP accuracy (less than null classification). The values of accuracy, sensitivity and specificity are shown in Table 4.

**Table 4.** Accuracy, sensitivity and specificity of Gatk and FreeBayes

| Method | Accuracy (%) | Sensitivity (%) | Specificity (%) |
|---|---|---|---|
| GATK | 77 | 70 | 93 |
| SNVer | 69 | 95 | 0 |
| FreeBayes | 60 | 44 | 100 |
| CRISP | 50 | 31 | 100 |

**Table 5.** Examples of positions associated to mutations according to the classification rules

| Position | Gene | Chromosome | T | G | C | A | Coverage | Pool |
|---|---|---|---|---|---|---|---|---|
| 152990560 | ABCD1 | X | 27 | 0 | 528 | 0 | 555 | #3 |
| 152990560 | ABCD1 | X | 52 | 0 | 968 | 1 | 1021 | #15 |
| 22301361 | ANO5 | 11 | 1132 | 13 | 0 | 29 | 1146 | #7 |
| 22301361 | ANO5 | 11 | 521 | 0 | 6 | 0 | 528 | #10 |

### 4.3  Decision Rules Results

The first step of the computational procedure consists in selecting only the positions characterized by a coverage greater than 500, one base with a frequency greater than 1.1 % and lower than 8 % and two bases below 1 %. Furthermore, since the replicated pools are composed by the same individuals of the original ones, each mutation, characterizing one sample, has to be replicated in 2 pools, one of the original group and one of the replicated group. For examples, two positions associated to possible mutations, according to the above rules, are reported in Table 5. In details, position 152990560, related to gene $ABCD1$ and chromosome $X$, is characterized by a coverage equal to 555, two bases, $G$ and $A$, below 1 % and base $T$ equal to 4.9 % in pool #3. In addition, this position is replicated in pool $S15$. An analogous situation is present in position 22301361 of gene $ANO5$ and chromosome 11. By means of the described filtering rules, 6502 mutations have been selected.

### 4.4  Standard Classifiers Performance

We check the accuracy of the classifier on a dataset consisting of 148 true mutations and 57 non-mutation positions (obtained by the biologists in a previous analysis). We applied a 10-fold cross-validation analysis using robust quantitative accuracy measurements of sensitivity and specificity on labelled training data. For this classifier we obtain an accuracy equal to 94.25 %, a sensitivity equal to 93 % and a specificity equal to 98 %. In order to check the adequacy of ReGEC algorithm, we compare different classifiers reported in Table 6. In details we consider: Support Vector Machines using Sequential Minimal Optimization (SMO) [22], Bayesian Network Classifier [23], $k$-Nearest Neighbours ($k$-NN) [24], Classification via Clustering (simple $k$-means), Multivariable functional

**Table 6.** Accuracy, sensitivity and specificity of different classification methods

| Method | Accuracy | Sensitivity | Specificity |
|---|---|---|---|
| ReGEC | 94.25 % ± 0.01 | 93 % ± 0.02 | 98 % ± 0.02 |
| SMO (linear) | 74.72 % ± 3.36 | 93 % ± 0.04 | 27 % ± 0.09 |
| Bayes Net | 92.82 % ± 3.14 | 93 % ± 0.04 | 92 % ± 0.06 |
| 5-NN | 85.61 % ± 4.43 | 90 % ± 0.05 | 73 % ± 0.11 |
| 10-NN | 81.34 % ± 4.31 | 90 % ± 0.04 | 60 % ± 0.11 |
| Classification via Clustering | 52.22 % ± 6.29 | 49 % ± 0.08 | 61 % ± 0.15 |
| RBFnetwork | 82.78 % ± 4.57 | 89 % ± 0.05 | 68 % ± 0.11 |
| Simple Logistic | 78.41 % ± 4.85) | 88 % ± 0.05) | 52 % ± 0.13 |
| Complement Naive Bayes | 65.33 % ± 5.50 | 62 % ± 0.07 | 75 % ± 0.10) |
| Bayesian Logistic Regression | 72.32 % ± 0.61 | 100 % ± 0.00 | 0 % ± 0.00 |
| Spegasos | 76.82 % ± 4.53 | 87 % ± 0.07 | 50 % ± 0.16 |

interpolation and adaptive networks [25], Simple Logistic [26,27], Complement Naive Bayes [28], Bayesian Logistic Regression [31] and Primal Estimated sub-GrAdient SOlver for SVM (Pegasos) [29].

As it is shown, ReGEC is better than the other considered classifiers in terms of accuracy, sensitivity and specificity.

### 4.5 Prediction

Using the set of 148 true mutations and 57 non-mutation positions as a training set, we test the 6502 positions obtained in the filtering step and we predict 1300 mutations. We consider only the loci corresponding to mutations in both pools. These predicted mutations were ranked using their distances from the plane of the belonging class. In details, we construct a rank as a ratio of the distance from the belonging class (+) to the distance from the non-belonging class (−):

$$rank = \frac{\frac{|w_+^T x_u - \gamma_+|}{\|w_+\|}}{\frac{|w_-^T x_u - \gamma_-|}{\|w_-\|}}.$$

The biologists start to check these mutations following the order of ranking and considering only exon positions. 114 mutations have been confirmed, including the control mutations, and 28 mutations have not been confirmed (false positive).

### 4.6 SIFT

In this section a structural analysis on the predicted mutations is conducted using SIFT (Sorting Intolerant From Tolerant). SNP studies identifies amino acid substitutions in protein-coding regions. Each substitution has the potential to affect protein function. SIFT is a program that predicts whether an amino acid substitution affects protein function. SIFT can distinguish between functionally neutral and deleterious amino acid changes in mutagenesis studies and on human polymorphisms (see, for more details, [30]). The result is that the mutations are 100 % damaging, that is, they are causative mutations.

## 5    Concluding Remarks

In this paper we consider pooled next generation sequencing data. We propose a computational procedure to identify and classify variants potentially related to human diseases. This procedure consists of two parts. The first step involves decision rules and in the second step we use a supervised classifier. We focus on causative mutations for muscular diseases and we compare different classification algorithms by means of a cross-validation framework.

In the near future, it will be interesting to consider the base qualities related to each base as features.

**Acknowledgment.** Authors would like to thank V. Nigro, M. Savarese, G. Di Fruscio, T. Giugliano, M. Iacomino, A. Torella, A. Garofalo, C. Pisano, F. Del Vecchio Blanco and G. Piluso (Seconda Universitá di Napoli, Patologia Generale), M. Mutarelli, V. Singh Marwah and M. Dionisi (TIGEM), and Italian LGMD network. This work has been partially funded by Italian Flagship project *Interomics* and by PON02_00619 projects.

## References

1. An Introduction to Next-Generation Sequencing Technology. www.illumina.com/ NGS
2. Licastro, D., Mutarelli, M., Peluso, I., Neveling, K., Wieskamp, N., Rispoli, R., Vozzi, D., Athanasakis, E., D'Eustacchio, A., Pizzo, M., D'Amico, F., Ziviello, C., Simonelli, F., Fabretto, A., Scheffer, H., Gasparini, P., Banfi, S., Nigro, V.: Molecular diagnosis of Usher syndrome: application of two different next generation sequencing-based procedures. PLoS ONE 7, Article number 43799 (2012)
3. Cacciottolo, M., Numitone, G., Aurino, S., Caserta, I.R., Fanin, M., Politano, L., Minetti, C., Ricci, E., Piluso, G., Angelini, C., Nigro, V.: Muscular dystrophy with marked dysferlin deficiency is consistently caused by primary dysferlin gene mutations. Eur. J. Hum. Genet. **19**, 974–980 (2011)
4. Nigro, V.: Improving the course of muscular dystrophy? (Editorial). Acta Myol. **31**, 109 (2012)
5. Kaplan, J.C.: The 2012 version of the gene table of monogenic neuromuscular disorders. Neuromuscul. Disord. **21**, 833–861 (2011)
6. Futschik, A., Schlotterer, C.: The next generation of molecular markers from massively parallel sequencing of pooled DNA samples. Genetics **186**, 207–218 (2010)
7. Calvo, S., Tucker, E., Compton, A., Kirby, D., Crawford, G., Burtt, N., Rivas, M., Guiducci, C., Bruno, D., Goldberger, O., Redman, M., Wiltshire, E., Wilson, C., Altshuler, D., Gabriel, S., Daly, M., Thorburn, D., Mootha, V.: High-throughput, pooled sequencing identifies mutations in NUBPL and FOXRED1 in human complex I deficiency. Nat. Genet. **42**(10), 851–860 (2011)

8. Wang, T., Pradhan, K., Ye, K., Wong, L.-J., Rohan, T.: Estimating allele frequency from next-generation sequencing of pooled mitochondrial DNA samples. Front. Genet. **2**, 51 (2011)
9. Ding, J., Bashashati, A., Roth, A., Oloumi, A., Tse, K., Zeng, T., Haffari, G., Hirst, M., Marra, M., Condon, A., Aparicio, S., Shah, S.: Feature-based classifiers for somatic mutation detection in tumour-normal paired sequencing data. Bioinformatics **28**, 167–175 (2012)
10. Next-Gen Sequencing: Advancing Sequencing for a Better World. Agilent Technologies Target Enrichment Solutions. www.agilent.com/genomics/ngs
11. Li, H., Handsaker, B., Wysoker, A., Fennell, T., Ruan, J., Homer, N., Marth, G., Abecasis, G., Durbin, R., G.P.D.P. Subgroup: The sequence alignment/Map format and SAMtools. Bioinformatics **25**, 2078–2079 (2009)
12. Mangasarian, O., Wild, E.: Multisurface proximal support vector machine classification via generalized eigenvalues. IEEE Trans. Pattern Anal. Mach. Intell. **28**, 69–74 (2006)
13. Parlett, B.N.: The Symmetric Eigenvalue Problem, p. 357. SIAM, Philadelphia (1998)
14. Wilkinson, J.H.: The Algebraic Eigenvalue Problem. Clarendon Press, Oxford (1988)
15. Guarracino, M.R., Cifarelli, C., Seref, O., Pardalos, P.: A classification algorithm based on generalized eigenvalue problems. Optim. Method Softw. **22**, 73–81 (2007)
16. Cifarelli, C., Guarracino, M., Seref, O., Cuciniello, S., Pardalos, P.: Incremental classification with generalized eigenvalues. J. Classif. **24**, 205–219 (2007)
17. DePristo, M.A., Banks, E., Poplin, R.E., Garimella, K.V., Maguire, J.R., Hartl, C., Philippakis, A.A., del Angel, G., Rivas, M.A., Hanna, M., McKenna, A., Fennell, T.J., Kernytsky, A.M., Sivachenko, A.Y., Cibulskis, K., Gabriel, S.B., Altshuler, D., Daly, M.J.: A framework for variation discovery and genotyping using nextgeneration DNA sequencing data. Nat. Genet. **43**, 491–498 (2011)
18. McKenna, A., Hanna, M., Banks, E., et al.: The genome analysis toolkit: a MapReduce framework for analyzing next-generation DNA sequencing data. Genome Res. **20**, 1297–1303 (2010)
19. Wei, Z., Wang, W., Hu, P., Lyon, G.J., Hakonarson, H.: SNVer: a statistical tool for variant calling in analysis of pooled or individual next-generation sequencing data. Nucleic Acids Res. **39**, 1–13 (2011)
20. Garrison, E., Marth, G. (2012). Haplotype-based variant detection from short-read sequencing. arXiv:1207.3907
21. Bansal, V.: A statistical method for the detection of variants from next-generation resequencing of DNA pools. Bioinformatics **26**, 318–324 (2010)
22. Platt, J.: Fast training of support vector machines using sequential minimal optimization. In: Schoelkopf, B., Burges, C., Smola, A. (eds.) Advances in Kernel Methods - Support Vector Learning. MIT Press, Cambridge (1998)
23. Friedman, N., Geiger, D., Goldszmidt, M.: Bayesian network classifiers. Mach. Learn. **29**, 131–163 (1997)
24. Fix, E., Hodges, J.L.: Discriminatory analysis, non parametric discrimination: consistency properties. Technical report 4, USAF School of Aviation Medicine, Randolph Field, Texas (1951)
25. Broomhead, D.S., Lowe, D.: Multivariable functional interpolation and adaptive networks. Complex Syst. **2**, 321–355 (1988)
26. Landwehr, N., Hall, M., Frank, E.: Logistic model trees. Mach. Learn. **95**, 161–205 (2005)

27. Sumner, M., Frank, E., Hall, M.: Speeding up logistic model tree induction. In: Jorge, A.M., Torgo, L., Brazdil, P.B., Camacho, R., Gama, J. (eds.) PKDD 2005. LNCS (LNAI), vol. 3721, pp. 675–683. Springer, Heidelberg (2005)
28. Rennie, J.D.M., Shih, L., Teevan, J., Karge, D.R.: Tackling the poor assumptions of naive bayes text classifiers. In: Proceedings of the Twentieth International Conference on Machine Learning, pp. 616–623 (2003)
29. Shalev-Shwartz, S., Singer, Y., Srebro, N.: Pegasos: primal estimated Sub-GrAdient SOlver for SVM. In: 24th International Conference on Machine Learning, pp. 807–814 (2007)
30. Ng, P.C., Henikoff, S.: SIFT: predicting amino acid changes that affect protein function. Nucleic Acids Res **31**, 3812–3814 (2003)
31. Mitchell, T.: Machine Learning. McGraw Hill, Berkshire (1997)

# A Framework for Mining Life Sciences Data on the Semantic Web in an Interactive, Graph-Based Environment

Artem Lysenko[1]([✉]), Jacek Grzebyta[1,2], Matthew M. Hindle[3],
Chris J. Rawlings[1], and Andrea Splendiani[1]([✉])

[1] Centre for Mathematical and Computational Biology, Rothamsted Research,
Harpenden, Herts AL5 2JQ, UK
{artem.lysenko,jacek.grzebyta,chris.rawlings,
andrea.splendiani}@rothamsted.ac.uk
[2] Structural and Molecular Biology, University College London,
London WC1E 6BT, UK
[3] SynthSys, University of Edinburgh, Edinburgh EH9 3JD, UK
matthew.hindle@ed.ac.uk

**Abstract.** The last decade saw the marked increase in the availability of
the Life Sciences data on the Semantic Web. At the same time, the need
to interactively explore complex and extensive biological datasets lead to
development of advanced visualisation tools, many of which present the
data in the form of a network graph. Semantic Web technologies offer
both a means to define rich semantics necessary to describe complex biological systems and allow large amounts of data to be shared effectively.
However, at present the need to be familiar with relevant technologies
greatly impedes access to these datasets by the non-specialist Life Sciences researches. To address this, we have developed a software framework that facilitates both access to the resources and presents the data
returned in an intuitive, graph-based format. Our framework is closely
integrated with Ondex, an established data integration solution in the
Life Sciences domain. The implementation consists of two parts. The first
one is a query console that allows expert users to execute Semantic Web
queries directly. The second one is a graph-based interactive browsing
solution that can be used to launch stock queries by choosing items in
the menu. In both cases, the result is re-formatted and visualised as a
graph in Ondex frontend.

## 1 Introduction

Compare to other domains of knowledge, Life Sciences have particularly complex
semantics, resulting from the convergence of different disciplines (e.g.: Medicine,
Genetics, Biochemistry, etc.) each with its proper language and conceptualiza-
tions. Reconciliation of these disparate data types is a pre-requisite for computationally driven analysis. Graphs (networks) model the data by representing
entities as nodes and relationships between them as edges. Graphs have been

© Springer International Publishing Switzerland 2014
E. Formenti et al. (Eds.): CIBB 2013, LNBI 8452, pp. 225–237, 2014.
DOI: 10.1007/978-3-319-09042-9_16

demonstrated to be both flexible and powerful modelling formalism for bringing together biological data and are increasingly popular among biologists [1]. In Life Sciences, the data relevant to particular problem is often spread out across multiple data providers [2] therefore the problem lies not only in the semantic reconciliation, but also in technical management of data storage, acquisition and convention between different formats. In this work we illustrate how both technical and semantic challenges of biological data integration can be addressed using graph-based modelling combined with federated data integration using Semantic Web technologies. Semantic Web is considered to be a particularly good fit for the data integration needs in a Life Sciences domain [3]. The Semantic Web formalism is based on an Resource Description Framework (RDF) format [4] for representing data. RDF can be queried using SPARQL query language [5], and web-based resources that can be queried in such way are called SPARQL endpoints. RDF format enforces the use of globally unique identifiers for all entities in the form of the Uniform Resource Identifiers (URIs) [6]. Therefore, on the Semantic Web all data is in part integrated by design: identifiers for the same entities are shared (or reconciled) across the entirety of the Web, and their properties inherit those semantics. In essence, data in RDF is represented as a set of triples (statements), each comprising subject, object and a predicate (property) that defines the association between the two. A subject is an RDF resource identified by a URI, an object can be either a resource or a literal (e.g. string, number, etc.) and a predicate is an instance of a type for that relationship. Although the type of a predicate can be a URI-identifiable resource, instances themselves cannot be assigned a URI and therefore cannot have any properties of their own. For this work, we have taken a well-established bioinformatics network analysis software Ondex [7], and developed an extension that allows it to access data on the Semantic Web. Ondex framework is made up of several tools including components for data integration, analysis, visualization and workflow composition and enactment [7]. The unifying feature across them is an Ondex graph data model - a Java-based implementation of graph formalism. What differentiates Ondex data model from those adopted by other graph-based bioinformatics tools, like BioTapestry [8] and Cytoscape [1], is that Ondex graph structure was specifically designed to facilitate data integration [9]. For this reason, Ondex graph is backed by a set of controlled vocabularies called Ondex metadata. Ondex metadata allows an unambiguous definition for types of nodes, edges and their attributes (Fig. 1). Because Ondex data model allows the definition of some semantics, in some ways it already resembles the Semantic Web model, even though there are also some important differences [10]. At the same time, Ondex data model was developed around entities of biological significance (e.g. proteins, genes, metabolic pathways) and is therefore more intuitive for the non-technical users compare to the relatively low-level RDF representation. This study focuses exclusively on the interactive graph visualisation component of the Ondex system called Ondex frontend. Ondex frontend presents the pre-integrated data to the user as an interactive and customisable network. It also provides other features expected in modern graph visualisation software, like

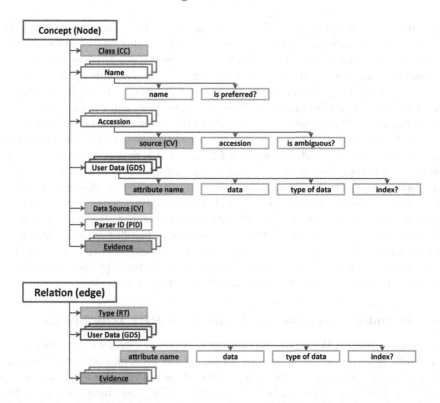

**Fig. 1.** Ondex data model, showing all possible attributes on Ondex nodes and edges. The shaded boxes indicate entities which are bound to the controlled vocabulary. The underline indicates where the presence of at least one attribute is compulsory and stacked boxes indicate where multiple attributes of that type are possible. Indentation is used to indicate sub-components of attributes composed of several fields

analysis-driven graph annotation (colour, size and shapes) as well as interactive inspection and modification of nodes and edges. This paper describes how Ondex data model was reconciled with RDF and an implementation of an interactive query solution for mining Life Sciences data on the Semantic Web.

## 2  Material and Methods

In this paper we propose a novel technical framework for exploring Life Sciences data on the Semantic Web. Our implementation also aims to provide a toolkit to make Semantic Web more accessible to the non-specialist users. To that end, we have also developed a versatile user interface via which non-experts can run pre-generated queries against Semantic Web resources retrieve context-specific RDF data without being exposed to the underlying Semantic Web technology. The defining feature of our approach is its ability to effectively combine data RDF and non-RDF data. Ondex already offers a rich set of import and integration options for dealing with non-RDF data. To support Semantic Web formalisms within the same system, we have developed a conversion strategy that can consistently reconcile Ondex data representation with RDF. It is worth pointing out that Ondex can already export data from its own format to RDF and this work primarily focuses on developing the capability to import RDF data. Although this export functionality will not be discussed here, the data conversion paradigms established during the development of Ondex exporter do have some consequences for the direction of this work. Specifically, the importer needs to be able to read in and correctly process Ondex-generated RDF. For this reason points about how Ondex data is represented in RDF will also be explained where relevant to the explanation of the technical requirements. To accommodate different levels of user expertise, our implementation consists of two distinct parts. The first one is a console that can be used to run SPARQL queries against specified endpoints. This is likely to be a method of choice for the users familiar with the Semantic Web and relevant resources. The second way of running queries is by selecting items from a context-sensitive menu, which allow context-specific execution of pre-generated quires without any technical expertise.

## 2.1    Semantic Reconciliation of Ondex and RDF Data Models

To import Semantic Web data into Ondex it is first necessary to establish a set of rules for interpreting RDF and re-packaging the data into an Ondex format. Although both formats can be loosely viewed as graphs, there are also some important semantic distinctions between them. The first distinction is in the representation of edges. In RDF an edge has a type and a direction but cannot have attributes. However, from the semantic point of view, addition of properties is possible if a statement is decomposed into its constituent subject, predicate and object entities (reification [4]). In that way, a predicate instance can become an RDF resource which can have properties of its own. In Ondex attributes are allowed and some are even required to be supplied at edge creation. To solve this, Ondex exporter relies on the reification, therefore an importer must be able to handle it as well. Consequently, at the import stage it is no longer possible to rely on an RDF resource always corresponding to a node, as it can also be part of a reified statement. To represent such data correctly it is necessary to clarify the identity of every resource. If a resource returned in response to user query is part of a reified statement, additional queries are needed to retrieve all other parts of the statement, so it can be correctly parsed. The developed implementation automates the launch of these additional queries to ensure that all reified statements are correctly interpreted as edges. The second distinction is the scope of the identifiers in Ondex and RDF formats. As RDF was designed as a data sharing format for the Web, it is inherently reliant on the unique identifiers for all resources. However, Ondex system was originally developed as a desktop application and therefore internally relies on identifiers that are only guaranteed to be unique within a context of one graph. In our implementation this was resolved by retaining an original URI for all imported resources as a special attribute. This URI then functions as a unique identifier when accessing the

Semantic Web. If a user executes a query that returns resources already present in the graph, they will be correctly reconciled and merged using their URIs. The third difference is in the verbosity of interpretation, e.g. how data is consolidated to create entities that are presented to the user. If interpreted literally, RDF can be represented as a graph of URIs and literals and typed edges. However, such a detailed, data-centric representation would be neither particularly useful, nor familiar to most users in the Life Science domain. In Life Sciences, entities normally seen as nodes in networks are often proteins, metabolites and genes, and relatively few of their attributes are visualised as links to other entities in the network, e.g. attributes such as database accessions, expression levels or sequences would not normally be represented as separate nodes in a network. Ondex network model is designed to cater to what is expected by the users from the Life Science community. Therefore, from the point of view of RDF such a representation is a higher-level abstraction of the actual data. This means that an Ondex node/edge would in fact correspond to an RDF sub-graph centred on a particular URI resource. To accommodate this, the importer launches additional queries to retrieve the declared types of all resources returned by user query. Based on these types, some of the resources are interpreted to create complex attributes of the Ondex data model instead of nodes. Additionally, all literal properties are always interpreted as attributes of Ondex nodes/edges.

## 2.2   SPARQL Query Console

The SPARQL console interface uses the functionality of the Apache Jena Semantic Web libraries (http://jena.apache.org) to forward queries entered by the user to a specified SPARQL endpoint. The console interface is completely reliant on the endpoint for processing the SPARQL query, and will therefore support all of the advanced SPARQL language features or even language extensions offered by the current endpoint. Additionally, the console offers two extra commands. The first is the connect command used to change the endpoint against which the queries are resolved. The second is use command to change the current configuration, a specially formatted flat file where additional options can be defined. An endpoint can return either data or an error message as response to a query. All error messages are displayed in the console, e.g. syntax or connection errors. For the cases where query returns a result, this result can either be an RDF graph (e.g. DESCRIBE and CONSTRUCT queries) or a table/value (e.g. ASK and SELECT queries). For the latter case, the results are formatted accordingly and displayed in a console window. For the former, returned RDF is passed to the on-the-fly parser that converts it into an Ondex graph. The parser will first execute any additional queries in a current configuration set. Then, it will apply the rules outlined in the previous section to interpret the combined result and create corresponding Ondex graph entities. The parser also retains an index of all URI attributes currently in an Ondex graph and will only create new entities if there is no matching entity in that index. An outline of the complete set of operations carried out to convert RDF to Ondex data model is shown in Fig. 2.

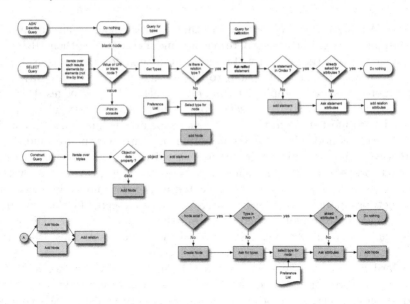

**Fig. 2.** A simplified schematic showing how RDF data is converted to fit into Ondex data model and queries for additional data triggered to populate required attributes

## 2.3   Interactive Browsing

Access to the interactive browsing was realised via a contextual popup menu, which can be opened by right-clicking on any visible nodes/edges in Ondex frontend. Each item in the menu is linked to a set of queries (Fig. 3), which are executed with respect to the URI of the graph element the menu applies to. As well as executing the queries directly linked to that menu items, each of them will also trigger any of the queries in the current configuration set – therefore, Ondex-specific parsing can be correctly applied to all results. Any actions done via a right-click menu are applied to all selected entities in the graph as well as an element that was clicked on to bring up the menu. The right-click menu is different for edges and nodes; therefore it is possible to specify different sets of queries appropriate to these elements. The data can be retrieved both from the endpoint or raw RDF file available at a particular URL on the Internet. If a menu entry is reading from the file, the URL is passed to the Jena library, which will download it and temporarily hold an internal representation of the data against which SPARQL queries will be executed. Such ability to access data in RDF files is particularly important to ensure that our implementation would benefit from the development of the Life Sciences resources supporting the Linked Data specification [11]. Briefly, the Linked Data rules dictate that a URI of the resource should be resolvable (i.e. lead to a resource on the Internet). If links resolve to a resource in RDF, Linked Data specification can be effectively used to navigate interlinked data across different resources. At present, this functionality is available in our implementation as a proof-of-concept only and

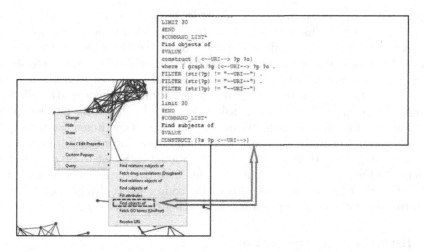

**Fig. 3.** An example screenshot showing an expanded SPARQL query sub-menu (bottom-left) and an extract from the interactive browsing menu configuration file (top-right) showing an entry for the find objects of item from the menu

will only resolve the UniProt URIs to populate their attributes. Queries launched via the right-click menu are processed by the same on-the-fly parser as those for the SPARQL console. From a users perspective, the experience is somewhat similar to browsing the web, whereby any node can be expanded by showing what other information is available about that entity in the current endpoint. The new graph elements created after the execution of the query can then be queried in the same manner, gradually building up an application-specific knowledge graph.

## 2.4   SPARQL Commands Configuration File

Both SPARQL console and interactive browsing menu can be associated with more than one query. By incorporating ability to chain executions of several queries, the system can to offer some intelligent re-interpretation of raw RDF data. The results can therefore be more accurately coerced into an Ondex data model, which is both more intuitive and more relevant to the way these data are customarily modelled in the Life Sciences. However, the use of query sets necessitated the development of additional mechanisms to link them i.e. it is necessary to define a way of passing the output from one query to the other(s). As in RDF all resources are identified by the URIs, this effectively means that the URIs from the result set need to be substituted into the right places in the following queries. This is done by defining appropriate place-holders that can be used in place of the URIs when constructing queries. At the execution time, the on-the-fly parser handles the replacement. A similar mechanism was used for defining queries in the interactive browsing menu except there, another set of place-holders was introduced that would be substituted for the URI of the node(s)/edge(s) selected by the user. In order to enable easy extensibility, these query sets are externalised in special flat-files. This allows the system to be easily re-targeted for use in a different application case without the need to change any of the application code. Only basic knowledge of SPARQL and understanding of how different URI place-holders are used are required to develop a new set of application-specific queries. The console only uses a configuration file that defines SPARQL prefixes and expansion queries. The browsing menu requires an additional file that binds query sets to particular menu items. Each entry in that file allows to specify the name of the menu item (including if it is intended for the edge or node right-click menu), an optional endpoint to be used and an actual set of queries. All of these files are updated live – e.g. the new configuration takes effect immediately, as soon as the source file is changed, without the need to restart the main application, facilitating rapid development.

## 3    Results and Discussion

As well as an application case outlined below, the SPARQL extension for Ondex frontend was tested against several RDF resources frequently used in bioinformatics; in particular we have used the UniProt [12], MyExperiment [13] and Bio2RDF [14] endpoints in our evaluation. For the set of general queries developed the performance was good enough to offer an interactive data exploration experience and it was possible to build up integrated networks for common bioinformatics analysis scenarios, like getting a set of proteins involved in particular biological processes and finding out what pathways they are involved in. We have developed an application case that approximates a common bioinformatics analysis scenario. The queries for this example were developed using Ondex SPARQL console and then adapted for use in the interactive browsing menu. We have included all of the commands developed for this example with the latest release of Ondex frontend software (available for download at www.ondex.org). Therefore, it should be possible to reproduce this example simply be selecting appropriate commands from the right-click menu, as described in the text below. Additionally, we have also included a set of generic commands that will fetch all available information about nodes and edges from the currently selected data source, which can be used to expand the network further.

### 3.1    Example Use Case Identifying Interacting Proteins with IPR002048 and IPR003527 Domains

To illustrate the developed method we present the following scenario. As a starting point we take a list of Arabidopsis thaliana protein accessions. The objective is to find out whether any of them encode proteins that both contain either IPR002048 (calcium binding) or IPR003527 (MAP kinase) domain and also are involved in a protein-protein interaction. As both calcium binding and kinase activity are commonly associated with signal transduction and regulation, an interaction between such proteins may be of significance to regulation of biological processes. This set-up is representative of a typical starting point for a bioinformatics analysis of experimental results. Many commonly-employed biological assays return lists of genes for follow-up study (e.g. microarray expression studies), which are then mined for meaningful patterns by consulting prior knowledge from various databases. The application case commands were added as items to the right-click interactive browsing menu. Each step was executed by selecting all relevant nodes in the graph and then launching appropriate query to fetch further information about them. In the queries shown, a~~URI~~token is replaced by a URI of a selected node at query execution. After a list of accessions is imported into Ondex, a node representing a protein is created for each of them, each having a URI attribute. In this example, the accessions we have started with were from Tair database [15], which does not offer its data on the Semantic Web. Therefore the first step was to find matching accessions from UniProt, by running a query against a UniProt Beta endpoint:

```
PREFIX : <http://purl.uniprot.org/core/>
CONSTRUCT {?protein rdfs:seeAlso <~~URI~~>}
WHERE {?protein rdfs:seeAlso <~~URI~~>}
```

This query creates an additional set of nodes representing protein entries from UniProt in the graph. As those nodes have UniProt URIs, which follow the LinkedData standard, it is possible to fetch additional information about these proteins by selecting them and choosing Resolve URL command from the menu. To access the protein-protein interaction data, we have chosen to use Bio2RDf endpoint. As Bio2RDF uses a different type of URIs to identify proteins, a query was run to build this mapping:

```
PREFIX : <http://purl.uniprot.org/core/>
CONSTRUCT {<~~URI~~> owl:sameAs ?bio2rdf}
WHERE {<~~URI~~> rdf:type :Protein .
BIND (URI(REPLACE(STR(<~~URI~~>), "http://purl.uniprot.org/uniprot/",
   "http://bio2rdf.org/uniprot:")) AS ?bio2rdf)}
```

The next task was to fetch all interaction for those proteins, by executing the following query against the Bio2RDF endpoint:

```
PREFIX ire: <http://bio2rdf.org/irefindex_vocabulary:>
PREFIX : <urn:default#>
CONSTRUCT {<~~URI~~> :participates_in ?s . ?b :participates_in ?s}
FROM http://bio2rdf.org/irefindex-all
WHERE {?s ire:interactor_a <~~URI~~> .?s ire:interactor_b ?b}
```

This query has returned 363 interactions (also represented by the nodes). Note that the query also returned all of the other proteins participating in those interactions and, where the match was found, they were correctly linked up to existing nodes. The last step is to find out which proteins have protein domains of interest. The domain information is brought in by running Get InterPro domains command from the menu that launches the following query against UniProt endpoint:

```
PREFIX : <http://purl.uniprot.org/core/>
CONSTRUCT {<~~URI~~> owl:hasValue ?resource}
WHERE {<~~URI~~> rdf:type :Protein ;rdfs:seeAlso ?resource .
FILTER REGEX (STR(?resource), "interpro", "i")}
```

The dataset produced as a result of executing all of these queries is shown in Fig. 4 (top). As the original objective was to find the proteins in the list that have a IPR002048 or IPR003527 domain and were also engaged in an interaction, an additional clean-up is required to actually focus in on the entities of interest. This can was done by selecting the two domain nodes and running a shortest path algorithm to remove all nodes that do not lie on a path between them. The result showed that there was one protein in the original set with an IPR003527 domain that is known to interact with 3 proteins possessing an IPR002048 domain (Fig. 4, bottom).

## 3.2  Discussion

This example presented above illustrates that the proposed method can be used effectively to locate relevant information on the Semantic Web and can be used to tackle certain classes of biological problems. However, at present our implementation does have several limitations, in particular with relation to the potentially unbounded nature of the Semantic Web. For example, some entities can have very large number of links and/or attributes, which may be unfeasible to import and visualise in their entirety. Currently, this is solved by restricting the number of returned results from queries that can potentially return large amounts of data. As this can lead to potentially missing some data, a better option would be to provide an interface that would summarise the results and allow the user to refine their query to reduce it to manageable number. The development of such an interface is one of the potential extensions to the system that will be considered in the future. As Ondex system also offers an extensive set of pre-existing parsers for other, non-RDF formats, including an ability to import simple tabular files, another potential application for this extension is combining the RDF-based and legacy data. As most of the URIs are currently constructed by simply combining the data provider URI with a database accession, the URI can often be generated if both accession and database is known. Interactive browsing menu made it possible for non-technical users of the Ondex system to benefit from the Life Sciences datasets now available on the Semantic Web. As each query is described in plain text in the menu and all data returned is presented in an intuitive way, this system can be used without requiring any knowledge of SPARQL or RDF. The envisaged usage for this functionality is to bring the federated data integration to the biological users of the Ondex system by changing the way bioinformatics and biologic researchers use Ondex for the collaborative research. At present, bioinformatics specialists usually prepare complex integrated datasets, which are then explored by biologists in Ondex frontend. Interactive browsing menu allows the integration step to be replaced

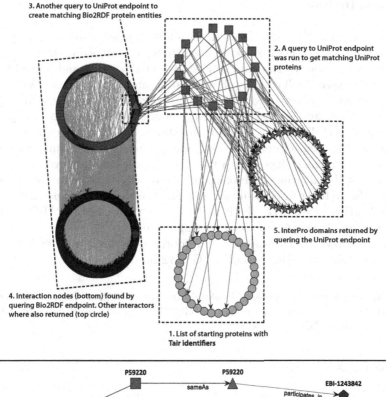

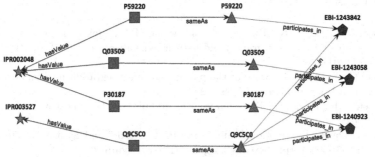

**Fig. 4.** Acquiring data using interactive query menu. An overview network showing the complete dataset and how it was constructed (top panel) and the results of applying the shortest path filter to show interactions between proteins possessing either IPR002048 or IPR003527 domain (bottom panel)

by a set of pre-defined queries relevant to the particular application case. Then, all integration can be done by the end-user, who would no longer be reliant on the bioinformatician for updates necessary to get access to the new data.

## 4    Conclusion

Using a Semantic-Web interface from within the Ondex front-end it is possible to interactively construct a knowledge graph, which is based exclusively around the entities of interest. The data can be inspected visually and refined further using an extensive toolkit of graph annotation and analysis tools. As only the data of relevance is retrieved, this strategy can be particularly advantageous when it is necessary to deal with large amounts of data and/or independent resources as very little data needs to be stored locally. Although other tools utilising a federated approach to data integration provide similar advantage, the solution presented does have a number of additional features that make it unique. First, by allowing additional queries to be specified in a flat-file by the end-user the system can be easily extended and re-shaped to fit particular needs and application cases. Secondly, it caters to both advanced (SPARQL console) and non-expert users (interactive browsing menu).

**Availability:** Ondex software and source code are available from www. ondex.org under the conditions of the GPLv2 licence.

**Acknowledgments.** Rothamsted Research receives grant in aid from the Biotechnology and Biological Sciences Research Council (BBSRC). This work was supported by the BBSRC award BBS/E/C/00005034.

## References

1. Smoot, M.E., Ono, K., Ruscheinski, J., Wang, P.L., Ideker, T.: Cytoscape 2.8: new features for data integration and network visualization. Bioinformatics **27**(3), 431–432 (2011)
2. Goble, C., Stevens, R.: State of the nation in data integration for bioinformatics. J. Biomed. Inform. **41**(5), 687–693 (2008)
3. Jenssen, T.K., Hovig, E.: The semantic web and biology. Drug Discov. Today **7**(19), 992 (2002)
4. W3C: Resource Description Framework (RDF) Model and Syntax Specification, vol. 2013 (1999). http://www.w3.org/TR/PR-rdf-syntax/
5. W3C: SPARQL Query Language for RDF, vol. 2013 (2008). http://www.w3.org/TR/rdf-sparql-query/
6. Berners-Lee, T.: RFC 3986 Uniform Resource Identifier (URI): Generic Syntax, vol. 2013 (2005). http://www.rfc-editor.org/rfc/rfc3986.txt
7. Kohler, J., Baumbach, J., Taubert, J., Specht, M., Skusa, A., Ruegg, A., Rawlings, C., Verrier, P., Philippi, S.: Graph-based analysis and visualization of experimental results with ondex. Bioinformatics **22**(11), 1383–1390 (2006)
8. Longabaugh, W.J.: Biotapestry: a tool to visualize the dynamic properties of gene regulatory networks. Meth. Mol. Biol. **786**, 359–394 (2012)
9. Taubert, J., Sieren, K., Hindle, M., Hoekman, B., Winnenburg, R., Philippi, S., Rawlings, C., Khler, J.: The oxl format for the exchange of integrated datasets. J. Integr. Bioinform. **4**(3), 62 (2007)

10. Splendiani, A., Rawlings, C.J., Kuo, S.-C., Stevens, R., Lord, P.: Lost in translation: data integration tools meet the semantic web (experiences from the ondex project). In: Gaol, F.L. (ed.) Recent Progress in DEIT, Vol. 2. LNEE, vol. 157, pp. 87–97. Springer, Heidelberg (2012)
11. Bizer, C., Heath, T., Berners-Lee, T.: Linked data - the story so far. Int. J. Seman. Web Inf. Sys. (IJSWIS) 5(3), 1–22 (2009)
12. Apweiler, R., Bairoch, A., Wu, C.H., Barker, W.C., Boeckmann, B., Ferro, S., Gasteiger, E., Huang, H., Lopez, R., Magrane, M., Martin, M.J., Natale, D.A., O'Donovan, C., Redaschi, N., Yeh, L.S.: Uniprot: the universal protein knowledgebase. Nucleic Acids Res. 32(Database issue), D115–D119 (2004)
13. Goble, C.A., Bhagat, J., Aleksejevs, S., Cruickshank, D., Michaelides, D., Newman, D., Borkum, M., Bechhofer, S., Roos, M., Li, P., De Roure, D.: myExperiment: a repository and social network for the sharing of bioinformatics workflows. Nucleic Acids Res. 38(Web Server issue), W677–W682 (2010)
14. Belleau, F., Nolin, M.A., Tourigny, N., Rigault, P., Morissette, J.: Bio2rdf: towards a mashup to build bioinformatics knowledge systems. J Biomed. Inform. 41(5), 706–716 (2008)
15. Rhee, S.Y., Beavis, W., Berardini, T.Z., Chen, G., Dixon, D., Doyle, A., Garcia-Hernandez, M., Huala, E., Lander, G., Montoya, M., Miller, N., Mueller, L.A., Mundodi, S., Reiser, L., Tacklind, J., Weems, D.C., Wu, Y., Xu, I., Yoo, D., Yoon, J., Zhang, P.: The arabidopsis information resource (tair): a model organism database providing a centralized, curated gateway to arabidopsis biology, research materials and community. Nucleic Acids Res. 31(1), 224–228 (2003)

# Combining Not-Proper ROC Curves and Hierarchical Clustering to Detect Differentially Expressed Genes in Microarray Experiments

Stefano Parodi[1], Vito Pistoia[2], and Marco Muselli[1(✉)]

[1] Institute of Electronics, Computer and Telecommunication Engineering, National Research Council of Italy, Via De Marini 6, 16149 Genoa, Italy
{parodi,muselli}@ieiit.cnr.it
[2] Laboratory of Oncology, G. Gaslini Children's Hospital, Largo G. Gaslini, 16147 Genoa, Italy
vitopistoia@ospedale-gaslini.ge.it

**Abstract.** *TNRC* (Test for Not Proper ROC Curve) is a statistical tool recently developed to identify differently expressed genes in microarray studies. In previous investigations it was demonstrated to be able to separate hidden subgroups in a two-class experiment, but being a univariate technique it could not exploit the complex multivariate correlation naturally occurring in gene expression data. In this study we show as the combination of *TNRC* with a standard technique of hierarchical clustering may provide useful biological insights. An example is provided using data from a publicly available data set of 4026 gene expression profiles in 42 samples of lymphomas and 14 samples of normal B cells.

**Keywords:** ROC analysis · Hierarchical clustering · Feature selection · Gene expression

## 1 Introduction

In the last four decades Receiver Operating Characteristic (ROC) curve analysis has been extensively used in biomedical setting for the evaluation of the performance of tumour markers for diagnostic and prognostic purposes [1–4]. In recent years, ROC curve parameters, including the whole and the partial area under the curve ($AUC$ and $pAUC$, respectively), have also been applied to feature selection tasks to identify potential markers from microarray experiments [5,6]. However, not-proper ROC curves that cross the ascending diagonal ("wiggly" curves) are in general discarded by standard methods of analysis, in that the value of $AUC$ and often also of $pAUC$ tend to be similar to those of a not informative curve [7].

Some statistical tests, able to identify wiggly ROC curves, have been developed, including two algorithms based on the projected length and on the area

E. Formenti et al. (Eds.): CIBB 2013, LNBI 8452, pp. 238–247, 2014.
DOI: 10.1007/978-3-319-09042-9_17

swept out by the ROC curve [7], Pietra and Gini indices of the corresponding Lorentz curve [8], and a test on the highest vertical distance between the rising diagonal and the curve. This latter can be estimated either on the whole set of observed values (in that case corresponding to the Kolmogorov-Smirnov statistics [4]) or an *a priori* identified range of specificity [9]. However, these methods are all unable to separate proper ROC curves from not-proper ones and also tend to show a low statistical power [9].

A statistical test for not-proper ROC curves (*TNRC*) has been recently developed, which was demonstrated to be able to identify differently expressed genes that tend to escape common statistical methods of feature selection [5,10]. *TNRC* is highly specific for not-proper curves and its statistical power clearly outperformed that of other statistical methods in a large simulation study [9]. Furthermore, differently from the above cited methods, a high level of the *TNRC* statistics can reveal hidden subgroups inside either one class under study [10]. In particular, *TNRC* was applied to a large dataset of gene expression profiles [11] and was able to identify 16 genes that had not been selected by two standard methods of analysis (namely, *AUC* and Student $t$ statistics). Interestingly, 13 out of the 16 corresponding not-proper ROC curves allowed to separate either the two hidden subclasses of malignant lymphomas (namely, CLL and FL) or the two hidden subgroups of differently stimulated normal cells [10].

A limit of the application of not-proper ROC curves is that it is impossible to assess if a high value of the *TNRC* statistics actually corresponds to hidden subclasses with clinical or biological meaning in the absence of some *a priori* information. In such case not-proper ROC analysis could take advantage from information derived from common methods of unsupervised data mining that have been largely applied in several biomedical fields including gene expression data analysis [12]. In the present investigation we will show how results from hierarchical clustering can contribute to the interpretation of not-proper ROC curves, comparing the expression profile of an apparently homogeneous group of diffuse large B-cell lymphoma (DLBCL) with that of a group of non-neoplastic B cells (NBC).

## 2    ROC Curve and the *TNRC* Statistics

Consider a sample of $n$ subjects, classified into two classes (A and B, respectively) on the basis of a binary outcome $Y$ taking values in $\{0, 1\}$. Suppose that a variable of interest (*e.g.*, the expression level of a given gene) is measured in all the $n$ individuals of the study. If $n_0$ is the number of subjects belonging to class A $(Y = 0)$, denote with $X_1, X_2, \ldots, X_{n_0}$ the values assumed by the variable of interest in this group of subjects, and denote with $W_1, W_2, \ldots, W_{n_1}$ the values measured in the $n_1$ individuals belonging to class B $(Y = 1)$. The empirical ROC curve can then be defined by considering different threshold values $c$ for the variable of interest and by computing the true and the false positive fractions, denoted by $TPF(c)$ and by $FPF(c)$, respectively, in the sample at hand [2,4]. It can be seen that:

$$TPF(c) = \frac{1}{n_1} \sum_{j=1}^{n_1} I(W_j \geq c), \quad FPF(c) = \frac{1}{n_0} \sum_{i=1}^{n_0} I(X_i \geq c) \qquad (1)$$

where $I$ is the indicator function providing $I(X_i = c) = 1$ if $X_i = c$ and $I(X_i = c) = 0$ otherwise. $TPF$ is often called the *sensitivity* of a diagnostic test, while $FPF$ corresponds to $1 - specificity$.

Let $AUC_k$ be the partial area under an ROC curve between the consecutive abscissa points $FPF(c_k-1)$ and $FPF(c_k)$, with $k = 1, \ldots, n$, computed according to the standard trapezoidal rule. The total area $AUC$ under the ROC curve is then given by:

$$AUC = \sum_{k=1}^{n_0} AUC_k = \sum_{k=1}^{n_0} \frac{1}{2}(TPF(c_k) + TPF(c_k - 1))(FPF(c_k) - FPF(c_k - 1))$$

When $TPF(c_k) = FPF(c_k)$ for any $k$, every threshold $c_k$ is not able to provide a valid classification for the two groups of subjects, *i.e.*, the class is assigned by chance. In this case we obtain a particular ROC curve, named the *chance line* (or *chance diagonal*) corresponding to the rising diagonal (Fig. 1, panel A). It should be observed that $AUC = 0.5$ for the chance line.

$AUC$ is strictly related to the Mann-Whitney $U$ statistics [13]. In particular, when referred to a gene expression profile, $AUC$ corresponds to the probability that a subject randomly selected from class B has a higher gene expression than a subject randomly selected from class A [14]. In most cases, the greater is the value of $AUC$, the higher is the difference between the two distributions [2,4]. Figure 1 shows an example of a proper (concave) ROC curve (panel A) derived from two normal distributions (panel B, plot I). However, in some cases the ROC curve is not-proper and crosses the chance line in one or more points (curve II in Fig. 1, panel A). In this case, even if the value of $AUC$ is close to 0.5, the two distributions can differ significantly (plot II in panel B). To recover these situations, the $TNRC$ statistics was introduced, by employing the following definition [10]:

$$TNRC = \sum_{k=1}^{n_0} |AUC_k - A_k| - |AUC - 0.5| \qquad (2)$$

where $A_k$ represents the partial area below the *chance line*.

When an ROC curve completely lies above (resp. below) the *chance line* we have $AUC_k \geq A_k$ (resp. $AUC_k < A_k$) for every $k = 0, 1, \ldots, n$, and (2) gives $TNRC = 0$. As a special case, this holds also for the *chance line*.

As shown in our previous paper [10], high values of $TNRC$ may correspond to a variety of not-proper ROC plots, including sigmoid and anti-sigmoid shaped curves. In particular, when a class of malignant cells samples is compared to non-neoplastic samples, considered as the referent (*i.e.*, corresponding to the class with $Y = 0$), sigmoid curves point out the presence of two hidden subclasses among normal cells, whereas anti-sigmoid curves indicate the presence of two hidden subclasses inside malignant cells. Finally, differently shaped not-proper

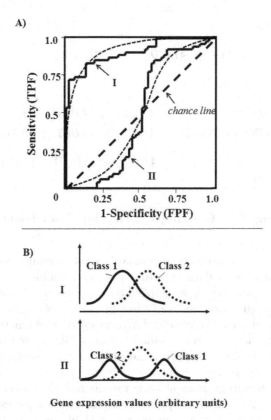

**Fig. 1.** Proper (concave, curve I) and not-proper (sigmoid, curve II) ROC curves (panel A) and the corresponding gene expression distributions (plot I and plot II, respectively, panel B).

curves can be occasionally observed. In general, they are difficult to interpret, and they may originate from multimodal distributions within either one class [10].

## 2.1   Properties of *TNRC*

It should be noted that the first part of the *TNRC* statistics in (2) corresponds to the area between the ROC curve and the chance diagonal ($ABCD$). Then, (2) can be rewritten as follows:

$$TNRC = ABCD - |AUC - 0.5|$$

Considering that $ABCD$ can be split into two subareas, namely the part above ($ABCD_a$) and below ($ABCD_b$) the *chance diagonal*, it can be easily shown that *TNRC* corresponds to the minimum value between $ABCD_a$ and $ABCD_b$:

$$TNRC = 2\min(ABCD_a, ABCD_b) \qquad (3)$$

As a matter of fact:

$$AUC = ABCD_a - ABCD_b + 0.5$$

that, replaced in (2), provides:

$$TNRC = ABCD - ABCD_a + ABCD_b = 2ABCD_b, \quad \text{if} AUC \geq 0.5$$
$$TNRC = ABCD + ABCD_a - ABCD_b = 2ABCD_a, \quad \text{if} AUC < 0.5$$

Since $ABCD_a \geq ABCD_b$ (resp. $ABCD_a < ABCD_b$) if $AUC \geq 0.5$ (resp. $AUC < 0.5$), (3) follows.

## 2.2   Interpreting *TNRC* Using Information from Hierarchical Clustering

Hierarchical clustering represents a standard simple unsupervised method for the analysis of microarray data able to exploit the complex correlation inside gene expression profiles [11]. When applied to an apparently homogeneous class, the associated plot (dendrogram) can identify subsets of samples belonging to hidden distinct subclasses. Conversely, *TNRC* is a supervised method that is also able to discover hidden clusters of samples, but, as illustrated above, it needs a referent group to make a comparison between the cumulative distributions of each feature in two classes.

Accordingly, a dendrogram identifying two distinct clusters can be combined with a not-proper ROC curve simply by merging the two corresponding plots, as it will be illustrated in the example reported in the Results section. The concordance between the hidden subclasses identified by the two methods can be assessed by standard statistical methods of bivariate analysis (*e.g.*, Pearson $\chi^2$ test, Fisher exact test or some index of concordance [15]).

A hierarchical clustering based on the Euclidean distance, was successfully applied to the large group of DLBCL considered for the present analysis (also including few samples from some selected lymphoma cell lines) and was able to identify two clusters with a signature characteristic of normal germinal cells (GC) and activated circulating B cells (AC), respectively [11]. Very interestingly, the two identified sub-classes corresponded to two groups of patients with a statistically significant different survival.

In order to combine information from *TNRC* and hierarchical clustering analysis we have classified our samples in over- and under-expressed on the basis of their location on the ROC curve, similarly to our previous investigation [10]. For this task, ROC curves identified by a high value of *TNRC* were separated into sigmoid shaped curves that can point out a bimodal distribution among NBC, and anti-sigmoid shaped curves, probably corresponding to a bimodal distribution among DLBCL samples. Other differently shaped curves were excluded from the analyses because considered as not-informative [10].

The two supposed hidden clusters were identified simply by splitting the ROC plot into two parts drawing a vertical line in the middle of the graph, thus crossing the x axis in the correspondence of 0.5 specificity. Samples lying at the left part of the graph were considered as over-expressed, whereas those lying at the opposite site were classified as under-expressed. It should be noted that, according to the above reported definition of *FPF*, in the presence of an anti-sigmoid curve, over-expressed (under-expressed) samples correspond to DLBCL with a gene expression higher (lower) than the median expression among NBC.

The association between over- and under-expression obtained from the not-proper ROC plots and the DLBCL classification from hierarchical clustering (namely, GC and AC) was assessed by the Pearson $\chi^2$ test, and *p*-values $<0.05$ were considered as statistically significant.

# 3  Results

Analysis was performed on a subset of samples from the database by Alizadeh *et al.* [11], which included 4026 gene expression profiles in many different samples of lymphomas or non-neoplastic cells. For the present analysis we selected a

**Table 1.** Comparison between the gene expression of 14 samples of normal circulating B cells and 42 samples of diffuse large B cell lymphomas by the *TNRC* test. The first 20 selected features, corresponding to the highest *TNRC* values, are listed.

| N | Gene ID | Gene name | *TNRC* | $P_{TNRC}$ | *AUC* |
|---|---------|-----------|--------|------------|-------|
| 1 | GENE1358X | *c-fos* | 0.1667 | 0.275 | 0.481 |
| 2 | GENE563X | *Similar proteasome sub. p112* | 0.1616 | 0.230 | 0.477 |
| 3 | GENE3494X | *ribonuclease 6 precursor* | 0.1553 | 0.170 | 0.511 |
| 4 | GENE3968X | *Deoxycytidylate deaminase* | 0.1536 | 0.220 | 0.490 |
| 5 | GENE289X | *Unknown* | 0.1514 | 0.330 | 0.453 |
| 6 | GENE789X | *KIAA0052* | 0.1446 | 0.135 | 0.487 |
| 7 | GENE813X | *Unknown* | 0.1429 | 0.340 | 0.438 |
| 8 | GENE790X | *Unknown* | 0.1417 | 0.210 | 0.532 |
| 9 | GENE173X | *Unknown* | 0.1406 | 0.075 | 0.489 |
| 10 | GENE1226X | *LD78 beta* | 0.1372 | 0.055 | 0.499 |
| 11 | GENE2474X | *Unknown* | 0.1366 | 0.100 | 0.498 |
| 12 | GENE1860X | *KSR1* | 0.1366 | 0.190 | 0.474 |
| 13 | GENE3493X | *ribonuclease 6 precursor* | 0.1361 | 0.140 | 0.533 |
| 14 | GENE1225X | *MIP-1 alpha* | 0.1338 | 0.160 | 0.537 |
| 15 | GENE1086X | *LYL-1* | 0.1321 | 0.175 | 0.520 |
| 16 | GENE2335X | *Unknown* | 0.1315 | 0.120 | 0.475 |
| 17 | GENE295X | *FBP1* | 0.1315 | 0.165 | 0.474 |
| 18 | GENE3967X | *Deoxycytidylate deaminase* | 0.1298 | 0.090 | 0.499 |
| 19 | GENE827X | *Unknown* | 0.1281 | 0.125 | 0.474 |
| 20 | GENE904X | *cote1* | 0.1253 | 0.065 | 0.503 |

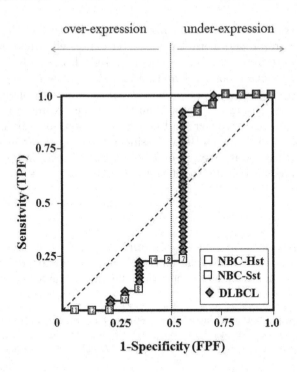

over-expression          under-expression

**Fig. 2.** Not-proper ROC curve corresponding to the expression of GENE1358X (c-fos) in Table 1. Hst = Highly stimulated; SSt = Slightly or not stimulated. NBC samples are numbered according to Alizadeh et al. (2000) [11].

class of 14 NBC, stimulated in different ways (6 heavily and 8 slightly or not stimulated) and a class of 42 DLBCL.

Feature selection was performed using the *TNRC* statistics. The first 20 genes corresponding to the highest *TNRC* values were retained. An estimate of the false discovery rate ($FDR$) was obtained from 200 permutations, using the method by Tusher *et al.* [16], while the probability for each gene to be included in the first 20 ones ($P_{TNRC}$ or $P_{AUC}$) was estimated by the method by Pepe *et al.* [6] using 200 bootstrapped samples.

Similarly to our previous analysis [10], selected genes were grouped on the basis of their function as follows: lymphocyte related genes (group 1), major histocompatibility complex related genes (group 2), genes involved in malignant cell transformation (group 3), genes related to nucleic acid metabolism or DNA transcription (group 4), and gene encoding various enzymes/kinases (group 5). In spite of some overlap, this classification allows to subdivide the tested genes according to their functional features.

Table 1 shows the results of the comparison between DLBCL and NBC. Seven genes had an unknown function at the time of the microarray experiment (genes n. 5, 7, 8, 9, 11, 16, 19), while the remaining fell in group 1 (genes n. 10, 14),

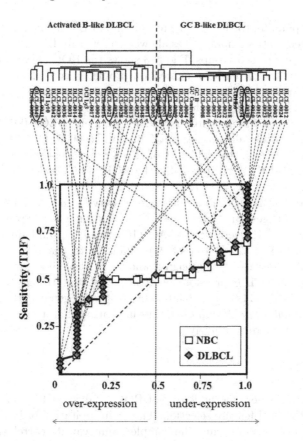

**Fig. 3.** Not-proper ROC curve corresponding to the expression of GENE3968X (Deoxy-cytidilate deaminase) in Table 1, and sample classification by hierarchical clustering. The corresponding dendrogram above the ROC curve was adapted from the original Fig. 3 in Alizadeh et al., 2000 [11] (http://www.nature.com/nature/journal/v403/n6769/full/403503a0.html), with permission.

group 3 (genes n. 15, 17), group 4 (genes n. 1, 3, 4, 6, 13, 18) or group 5 (genes n. 2 and 12).

$AUC$ estimates were all close to the expected value of 0.5 and, accordingly, the corresponding selection probabilities equal 0 for any comparison (not shown in Table 1). Conversely, values of $TNRC$ statistics ranged from 0.1253 to 0.1667 and the corresponding selection probability $P_{TNRC}$ varied between 0.065 to 0.275. $FDR$ estimate was 16.9 %. Anti-sigmoid shaped curves were observed in five cases (genes n. 4, 11, 12, 15 and 18), whereas the remaining curves all had a rather regular sigmoid shape, which indicated the existence of at least two hidden subclasses within the NBC class. As an example, ROC curve for GENE1358X (*c-fos*) is shown in Fig. 2. All the six samples lying in the right side of the plot corresponded to highly stimulated B cells, while the remaining eight samples, corresponding to slightly or not stimulated B cells, lied at the opposite side. As expected, anti-sigmoid curves allowed to separate DLBCL in (allegedly) over- and under-expressed groups. Figure 3 shows the ROC curve corresponding to the *Deoxycytidilate deaminase* expression profile (gene n. 4 in Table 1). The large majority of samples in the right part of the curve corresponded to GC B-like DLBCL identified by the hierarchical clustering in the original paper by Alizadeh *et al.* [11], with only three exceptions (namely: DLCL-005, DLCL-0021 and DLCL-0049), while all samples but four (DLCL-0030, DLCL-0034, DLCL-0020, and DLCL-0051) in the left part of the curves (over-expressed respect to NBC) corresponded to Activated B-like DLBCL. This association was highly statistically significant ($\chi^2 = 18.7, p < 0.001$).

With regards to the other four anti-sigmoid curves, no association was found between gene n. 11, gene n. 12 and the GC/AC status. Over-expressed DLBCL samples for gene n. 15 were mostly AC (16 out of 21) and under-expressed samples were mostly GC (11 out of 21), similarly to that observed for gene n. 4, but in this case statistically significance was borderline ($\chi^2 = 3.635, p = 0.057$). Finally, DLBCL expression for gene n. 18 was strongly associated to GC/AC status ($\chi^2 = 11.96, p = 0.001$). Interestingly, gene n. 18 is a clone of gene n. 4 (*Deoxycytidilate deaminase*), thus indicating that a chance finding due to multiple testing is very unlikely.

## 4    Conclusions

*TNRC* represents a new methodology of ROC analysis, which belongs to the supervised methods of feature selection. The main limitation of ROC analysis is that it cannot take into account the complex multivariate correlation between features that is commonly encountered in gene expression databases. Conversely, hierarchical clustering is an unsupervised methodology that can identify hidden subgroups of genes and/or samples exploiting the distance between features in a multivariate Euclidean space [17]. The main limit of this technique is the tendency to find pseudo-clusters also in data sets of randomly generated features. Results from the present investigation, even if still explorative, indicates that the combination of not-proper ROC analysis with traditional hierarchical clustering can provide useful insights for the interpretation of gene expression data.

## References

1. Kampfrath, T., Levinson, S.S.: Brief critical review: statistical assessment of biomarker performance. Clin. Chim. Acta **419**, 102–107 (2013)
2. Alemayehu, D., Zou, H.: Applications of ROC analysis in medical research: recent developments and future directions. Acad. Radiol. **19**, 1457–1464 (2012)
3. Parodi, S., Muselli, M., Carlini, B., Fontana, V., Haupt, R., Pistoia, V., Corrias, M.V.: Restricted ROC curves are useful tools to evaluate the performance of tumour markers. Stat. Methods Med. Res. 26 Jun 2012 [epub ahead of print]
4. Pepe, M.: The Statistical Evaluation of Medical Tests for Classification and Prediction. Oxford University Press, Oxford (2003)

 5. Silva-Fortes, C., Turkman, M.A., Sousa, L.: Arrow plot: a new graphical tool for selecting up and down regulated genes and genes differentially expressed on sample subgroups. BMC Bioinf. **13**, 147 (2012)
 6. Pepe, M., et al.: Selecting differentially expressed genes from microarray experiments. Biometrics **59**, 133–142 (2003)
 7. Lee, W., Hsiao, C.: Alternative summary indices for the receiver operating characteristic curve. Epidemiology **7**, 605–611 (1996)
 8. Lee, W.: Probabilistic analysis of global performances of diagnostic tests: interpreting the Lorentz curve-based summary measures. Stat. Med. **18**, 455–471 (1999)
 9. Kagaris, D., Yiannoutsos, C.: A multi-index ROC based methodology for high throughput experiments in gene discovery. Int. J. Data Min. Bioinf. **8**, 42–65 (2013)
10. Parodi, S., Pistoia, V., Muselli, M.: Not proper ROC curves as new tool for the analysis of differentially expressed genes in microarray experiments. BMC Bioinf. **9**, 410 (2008)
11. Alizadeh, A., et al.: Distinct types of diffuse large B-cell lymphoma identified by gene. Nature **403**, 503–511 (2000)
12. Michiels, S., Kramar, A., Koscielny, S.: Multidimensionality of microarrays: statistical challenges and (im)possible solutions. Mol. Oncol. **5**, 190–196 (2011)
13. Bamber, D.: The area above the ordinal dominance graph and the area below the receiver operating characteristic graph. J. Math. Psychol. **12**, 387–415 (1975)
14. Parodi, S., Muselli, M., Fontana, V., Bonassi, S.: ROC curves are a suitable and flexible tool for the analysis of gene expression profiles. Cytogenet. Genome Res. **101**, 90–91 (2003)
15. Gibbons, J.D., Chakraborti, S.: Nonparametric Statistical Inference, 4th edn. Marcel Dekker Inc, New York (2003)
16. Tusher, V., Tibshirani, R., Chu, G.: Significance analysis of microarrays applied to the ionizing radiation response. Proc. Nat. Acad. Sci. USA **98**, 5116–5121 (2001)
17. Hastie, T., Tibshirani, R., Friedman, J.: Hierarchical clustering. The Elements of Statistical Learning, 2nd edn, pp. 520–528. Springer, New York (2009)

# Fast and Parallel Algorithm for Population-Based Segmentation of Copy-Number Profiles

Guillem Rigaill[1]([envelope]), Vincent Miele[2], and Franck Picard[2]

[1] Unité de Recherche en Génomique Végétale (URGV) INRA-CNRS-Université d'Evry Val d'Essonne, 2 Rue Gaston Crémieux, 91057 Evry Cedex, France
`rigaill@evry.inra.fr`
[2] Laboratoire de Biométrie et Biologie Evolutive,
UMR CNRS 5558 Université Lyon 1, 69622 Villeurbanne, France
`{vincent.miele,franck.picard}@univ-lyon1.fr`

**Abstract.** Dynamic Programming (DP) based change-point methods have shown very good statistical performance on DNA copy number analysis. However, the quadratic algorithmic complexity of DP has limited their use on high-density arrays or next generation sequencing data. This complexity issue is particularly critical for segmentation and calling of segments, and for the joint segmentation of many different profiles. Our contribution is two-fold. First we provide an at worst linear DP algorithm for segmentation and calling, which allows the use of DP-based segmentation on high-density arrays with a considerably reduced computational cost. For the joint segmentation issue we provide a parallel version of the `cghseg` package which now allows us to analyze more than 1,000 profiles of length 100,000 within a few hours. Therefore our method and software package are adapted to the next generation of computers (multi-cores) and experiments (very large profiles).

**Keywords:** DNA copy number · Dynamic Programming · Segmentation · Joint segmentation · Parallel computing

## 1 Introduction

Segmenting heterogeneous signals into regions of common characteristics is often required for biologists that face high dimensional information. This partitioning helps reducing the dimension of the data and provides guidelines for interpretation and further biological investigation. Such methods have been widely applied in Genomics, to unravel DNA sequences structures using base composition [6], to segment expression profiles [2,7], and to determine copy-number variations based on array CGH data. Microarray CGH data analysis is certainly the field for which every possible method of segmentation have been tried. Three main categories of methods lead to many developments: Circular Binary Segmentation (CBS, [8]), Hidden Markov Models [5], and change-point analysis [10,11]. Our purpose

© Springer International Publishing Switzerland 2014
E. Formenti et al. (Eds.): CIBB 2013, LNBI 8452, pp. 248–258, 2014.
DOI: 10.1007/978-3-319-09042-9_18

in this work is not to compare the different methods for array CGH analysis. Such comparisons have been done elsewhere [18], and extensive reviews now exist on the subject [14,16]. Our focus is the following: in every comparative study, change-point analysis based on Dynamic Programming (DP) has shown excellent performance along with CBS [3,18]. The advantage of DP-based segmentation is that it provides the best global segmentation, i.e. the segmentation that globally minimizes a likelihood criterion, whereas local approaches like CBS only provide local minimizers. Moreover, change-point models can also integrate a calling step that is used to cluster segments that show the same copy-number on average (also called segmentation/clustering) [11,17], which has been shown to be of central importance in the segmentation process [18]. Unfortunately, the algorithmic complexity of DP is proportional to the square of the signal's length, which has hampered the use of such method on high-density arrays for instance. This issue has become even more problematic when dealing with population-based or joint segmentation [9,15,20]. Our contribution is two-fold: first we provide a linearized version of the DP algorithm for segmentation/clustering adapting pruning strategies that have recently emerged in the field [4,13]. Secondly we deal with the joint segmentation issue by providing a parallel version of existing algorithms implemented in the cghseg package. With the growing availability of multicore computers (from laptops to many-core servers), it has become essential to provide software that use every available resource. R has been a tremendous platform for package distribution, and here we provide a new version of the cghseg package, a *next generation package* that is adaptive to available computing power. The performance of cghseg are impressive: segmenting 1,000 profiles of length 100,000 can now be done in few hours, which was impossible before. This makes segmentation models a new exact investigation method that can be used in routine for exploratory as well as deep analysis.

## 2 Linearization of Dynamic Programming for Segmentation and Segmentation/Clustering

The purpose of segmentation is to partition a signal of $n$ observations $\{Y(t)\}$ into $K$ segments of homogeneous distributional parameter. In this section we deal with the univariate segmentation of one array CGH profile, the case of Joint multivariate segmentation being considered in Sect. 3. In the following a segment is an interval delimited by two change-points $\tau_k, \tau_{k+1}$ for instance. $r_k = [\![\tau_k, \tau_{k+1} - 1]\!]$ stands for segment $k$. To stick to this definition for the first and last segment we use the convention $\tau_0 = 1$ and $\tau_K = n + 1$. A segmentation in $K$ segments is denoted by $m^{(K)} = \{r_0, r_1, \ldots, r_{K-1}\}$.

A standard statistical model for segmentation is the detection of changes in the mean of a signal, such that $\forall t \in r_k,\ Y(t) \sim \mathcal{N}(\mu_k, \sigma^2)$. In the special case of array CGH, an additional step is the "calling" step that is performed by introducing additional (hidden) label variables $\{Z(r_k)\}_k$ to cluster each segment into categories such as `deleted`, `normal`, `amplified` (not limited to 3 states). Then the segmentation/clustering model becomes:

$$\forall t \in r_k, \ \forall p \in \{1...P\}, \ Y(t)|\{Z(r_k) = p\} \sim \mathcal{N}(\mu_p, \sigma^2),$$

with $P$ the total (fixed) number of hidden states, with $\pi_p$ the proportion of segments in each state.

Once the statistical model has been defined, the main algorithmic challenge lies in the exact determination of the boundaries of segments $\{\tau_k\}$ (and not in the estimation of mean parameters $\mu_k$s or $\mu_p$s depending on the model). A well known solution to this problem is to use Dynamic Programming for a given number of segments $K$ to find the best *global* segmentation in terms of "cost" (to be defined). In this work, we do not deal with the issue of model selection to estimate $K$ and $P$, discussed elsewhere [10, 19]. To perform Dynamic Programming, we need to define the "unit" cost of a generic segment $r = [\![t_1, t_2]\!]$, which is given by minus the local log-likelihood calculated on $r$:

$$C_{(r)}^{(1)} = \begin{cases} \sum_{t \in r}(y(t) - \mu_r)^2/2\sigma^2, \text{ for segmentation in the mean} \\ -\log\left(\sum_p \pi_p \exp\left\{-\sum_{t \in r}(y(t) - \mu_p)^2/2\sigma^2\right\}\right), \text{ for seg/clust,} \end{cases}$$

with superscript (1) standing for "one segment". A main difference between the two models lies in the estimation of the mean parameters. In the case of segmentation in the mean, parameters $\{\mu_k\}_k$ can be estimated directly by the empirical means of segments while computing the position of the breaks. In the case of segmentation/clustering, parameters $\{\mu_p\}_p$ are common across segments. Consequently, they are fixed while computing breakpoint coordinates, and they are estimated iteratively by using an EM-algorithm, leading to a so-called DP-EM algorithm [11]. In the case of segmentation/clustering, we propose to simplify the cost function by using a *classification cost function*, an approximation denoted by $\widetilde{C}_{(r)}^{(1)}$ which consists in focusing on the dominant term within the sum over $P$ exponentials:

$$\widetilde{C}_{(r)}^{(1)} = \min_p \left\{ \sum_{t \in r} \frac{(y(t) - \mu_p)^2}{2\sigma^2} + \log(\pi_p) \right\}. \tag{1}$$

This cost function is analog to the cost function that is used in standard k-means algorithms.

Since the purpose is to find the global minimum of the total cost function into $K$ segments, we also introduce the set of all segmentations of a given segment $r$ into $K$ segments such that $\mathcal{M}_{(r)}^{(K)} = \mathcal{M}_{t_1,t_2}^{(K)}$. Then the optimal cost of a segmentation of $r$ into $K$ segments and its associated optimal segmentation are defined as:

$$C_{(r)}^{(K)} = \min_{m \in \mathcal{M}_{(r)}^{(K)}} \left\{ \sum_{r \in m} C_{(r)}^{(1)} \right\}, \text{ and } \widehat{m}_{(r)}^{(K)} = \operatorname*{argmin}_{m \in \mathcal{M}_{(r)}^{(K)}} \left\{ \sum_{r \in m} C_{(r)}^{(1)} \right\}.$$

Similarly, when Approximation $\widetilde{C}_{(r)}^{(1)}$ is used (Eq. (1)), we use notations $\widetilde{C}_{(r)}^{(K)}$ and $\widetilde{m}_{(r)}^{(K)}$. When the cost of a segmentation is segment additive (which is the case in both models), a $\mathcal{O}(Kn^2)$ Dynamic Programming algorithm can be built to recover the best exact segmentation (into 1 to $K$ segments).

## 2.1 Original Dynamic Programming Algorithm for Segmentation

A basic statement is that the cost of a given segmentation is the sum of the cost of its segments. Thus the Bellman optimality principal holds and we have: $C_{1,t}^{(k+1)} = \min_{\tau \le t}\{C_{1,\tau-1}^{(k)} + C_{\tau,t}^{(1)}\}$. Using this update rule a Dynamic Programming algorithm can be built to recover all $C_{1,t}^{(k+1)}$ for all $t \le n$ and $k \le K$. This can be done using Algorithm 1 for instance. For simplicity we did not include the initialization of all $C_{1,t}^{(1)}$ for $t \le n$ and of all $C_{1,t}^{(k)}$ for $k \le K$ and $t < k$. All $C_{1,t}^{(k)}$ are initialized as $+\infty$ and $C_{1,t}^{(1)}$ are initialized using their definition. This algorithm assumes that all $C_{t_1,t_2}^{(1)}$ have been pre-computed and stored (in a $n$ by $n$ matrix) or that they can be efficiently computed on the fly, which is the case for every models we consider here. At step $k,t$ of Algorithm 1 $\mathcal{O}(t)$ basic operations are performed. If we sum these for all $k < K$ and $t < n$ we see that the algorithm has a $\mathcal{O}(Kn^2)$ time complexity. This $n^2$ factor is the main reason why Dynamic Programming can be prohibitive to use on large signals (like SNP arrays for instance).

---

**Algorithm 1.** Standard DP algorithm

---

**Input:** $Y(t)$ a profile of $n$ observations, $K$ an integer
**Output:** $C_{1,t}^{(k)}$ in $\mathbb{R}$ and $M_{1,t}^{(k)}$ in $\mathbb{N}$ for all $k \le K$ and $t \le n$
**for** $t \in [\![1,n]\!]$ **do**
    $C_{1,t}^{(1)} = C_{1t}$ ; $M_{1,t}^{(1)} = 0$
**end for**
**for** $k \in [\![2, \min(t, K)]\!]$ **do**
    **for** $t \in [\![1,n]\!]$ **do**
        $C_{1,t}^{(k)} = \min_{k-1 \le \tau \le t-1}\{C_{1,\tau}^{(k-1)} + C_{(\tau+1)t}^{(1)}\}$ ; $M_{1,t}^{(k)} = \operatorname*{argmin}_{k-1 \le \tau \le t-1}\{C_{1,\tau}^{(k-1)} + C_{(\tau+1)t}^{(1)}\}$
    **end for**
**end for**

---

## 2.2 A Linear Dynamic Programming Algorithm for the Classification Cost Function

An important consequence of using the classification cost $\widetilde{C}_{(r)}^{(1)}$ rather than $C_{(r)}^{(1)}$ is that the set of candidate segmentations can be pruned efficiently. The idea of pruning the set of candidate segmentations is not new and was proposed for other cost functions [4,13]. In these two algorithms the pruning step usually allows for an important speed up and the average time complexity is for many signal in $\mathcal{O}(n)$ or $\mathcal{O}(n\log(n))$. Nonetheless, in both cases, the worst case is quadratic with respect to $n$. In the case of the classification cost however the pruning step is particularly efficient and we can guarantee that the time and space complexity of the algorithm are at worst in $\mathcal{O}(KPn)$. Furthermore, we can also guarantee that the average cost (over all data points) of the recovered segmentation $\widetilde{m}_{(r)}^{(K)}$ is at worst within $K \log(P)/n$ of the optimal segmentation $\widehat{m}_{(r)}^{(K)}$ (see Theorem 22).

Before we describe the algorithm we need to define some new notations. We define the approximate cost of a segment knowing its mean $\mu$ as $\widetilde{C}_{(r)}^{(1)}(\mu) = \left\{\sum_{t \in r} (y(t) - \mu)^2/2\sigma^2 - \log(\pi_p)\right\}$. Using this notation we can rewrite the classification cost of a segment $r = [t_1, t_2]$ as $\widetilde{C}_{t_1,t_2}^{(1)} = \min_p\{\widetilde{C}_{t_1,t_2}^{(1)}(\mu_p)\}$, with $\widetilde{C}_{t_1,t_2}^{(1)}(\mu)$ being point additive in the sense that $\widetilde{C}_{t_1,t_2}^{(1)}(\mu) = \sum_{t_1 \le t \le t_2} \widetilde{C}_{t,t}^{(1)}(\mu)$. Then we define the cost of the best segmentation knowing that the mean of the last segment is $\mu$ as:

$$\widetilde{C}_{1,t}^{(K)}(\mu) = \min_{m \in \mathcal{M}_{(1,t)}^{(K)}} \left\{ \sum_{k < K-1} \widetilde{C}_{(r_k)}^{(1)} + \widetilde{C}_{(r_K)}^{(1)}(\mu) \right\}.$$

Using this notation we get that $\widetilde{C}_{1,t}^{(k)} = \min_{p < P} \left\{ \widetilde{C}_{1,t}^{(k)}(\mu_p) \right\}$, and if we know every $C_{1,t}^{(k)}(\mu_p)$ at step $t$ for all $p < P$, we straightfowardly get $C_{1,t}^{(k)}$ in $\mathcal{O}(p)$. As $\widetilde{C}_{t_1,t_2}(\mu)$ is point additive updating $\widetilde{C}_{t_1,t_2}(\mu)$ is easy and can be done efficiently using the following theorem.

**Theorem 21.** $\widetilde{C}_{1,t+1}^{(k)}(\mu) = \min\left\{ \widetilde{C}_{1,t}^{(k)}(\mu), \widetilde{C}_{1,t}^{(k-1)} \right\} + \widetilde{C}_{t+1,t+1}^{(1)}(\mu)$

*Proof.* Let us first notice that: $\widetilde{C}_{1,t+1}^{(k)}(\mu) = \min_{\tau < t+1} \left\{ \widetilde{C}_{(1,\tau)}^{(k-1)} + \widetilde{C}_{(\tau+1,t)}^{(1)}(\mu) \right\}$. Using the definition of $\widetilde{C}_{1,t}^{(k)}(\mu)$ we get that:

$$\widetilde{C}_{1,t}^{(k)}(\mu) + \widetilde{C}_{(t+1,t+1)}^{(1)}(\mu) = \min_{\tau < t} \left( \widetilde{C}_{(1,\tau)}^{(k-1)} \right) + \widetilde{C}_{(t+1,t+1)}^{(1)}(\mu)$$

From this the theorem follows.     ∎

Using this theorem, knowing $\widetilde{C}_{1,t}^{(k)}(\mu)$ and $\widetilde{C}_{1,t}^{(k-1)}$ we get $\widetilde{C}_{1,t+1}^{(k)}(\mu)$ in $\mathcal{O}(1)$ and we derive Algorithm 2 for the DP step of the DP-EM algorithm [11]. For simplicity we did not include the initialization of $\widetilde{C}_{1,t}^{(1)}$ for $t < n$ and $\widetilde{C}_{1,1}^{(1)}(\mu_p)$. All $\widetilde{C}_{1,1}^{(k)}(\mu_p)$ are initialized as $+\infty$ and $\widetilde{C}_{1,t}^{(1)}$ are initialized using their definition. At step $k, t$ of Algorithm 2 $\mathcal{O}(P)$ basic operations are performed. If we sum these for all $k < K$ and $t < n$ we straightforwardly see that the algorithm has an $\mathcal{O}(KPn)$ time complexity.

## 2.3   A Bound on the Quality of the Approximation

Using the approximation defined in Eq. 1 we can guarantee the quality of the obtained segmentation using the following theorem.

---

**Algorithm 2.** Linear DP algorithm for the classification cost

---

**Input:** $Y(t)$ a profile of $n$ observations, $K$ an integer
**Input:** $C_{t_1,t_2}$ cost of the segments $]t_1, t_2]$ for all $(t_1, t_2) \in [\![1,n]\!]^2$
**Output:** $C_{1,t}^{(k)}$ in $\mathbb{R}$ and $M_{1,t}^{(k)}$ in $\mathbb{N}$ for all $k \leq K$ and $t \leq n$
for $k \in [\![2, \min(t, K)]\!]$ do
    for $t \in [\![1, n]\!]$ do
        for $p \in [\![1, P]\!]$ do
            $C_{1,t}^{(k)}(\mu_p) = \min\{C_{1,t-1}^{(k)}(\mu_p), C_{1,t-1}^{(k-1)}\} + C_{t,t}^{(1)}(\mu_p)$ ;
            $M_{1,t}^{(k)}(\mu_p) = \text{argmin}\{C_{1,t-1}^{(k)}(\mu_p), C_{1,t-1}^{(k-1)}\}$
        end for
        $C_{1,t}^{(k)} = \min_p\{C_{1,t}^{(k)}(\mu_p)\}$ ; $p^* = \text{argmin}_p\{C_{1,t}^{(k)}(\mu_p)\}$ ; $M_{1,t}^{(k)} = M_{1,t}^{(k)}(\mu_{p^*})$
    end for
end for

---

**Theorem 22.** *Using approximation defined in Eq. 1 we have for all segments $R$ and $K$*

$$\widetilde{C}_{(R)}^{(K)} - K\log(P) \leq \sum_{r \in \widehat{m}_{(R)}^{(K)}} \widetilde{C}_{(R)}^{(K)} - K\log(P) \leq C_{(R)}^{(K)} \leq \sum_{r \in \widetilde{m}_{(R)}^{(K)}} C_{(R)}^{(K)} \leq \widetilde{C}_{(R)}^{(K)}.$$

*Proof.* We have $C_{(r)}^{(1)} = -\log(\sum_p \exp\{-\widetilde{C}_{(r)}^{(1)}(\mu_p)\})$ and $\widetilde{C}_{(r)}^{(1)} \geq \widetilde{C}_{(r)}^{(1)}(\mu_p)$. From this we get that: $\forall r, \widetilde{C}_{(r)}^{(1)} - \log(P) \leq C_{(r)}^{(1)} \leq \widetilde{C}_{(r)}^{(1)}$, which gives, along with the definition of $C_{(R)}^{(K)}$:

$$\forall m \in \mathcal{M}_{(R)}^{(K)}, \widetilde{C}_{(R)}^{(K)} - K\log(P) \leq \sum_{r \in m} \widetilde{C}_{(r)}^{(1)} - K\log(P) \leq \sum_{r \in m} C_{(r)}^{(1)}. \quad (2)$$

Similarly we get:

$$\forall m \in \mathcal{M}_{(R)}^{(K)}, C_{(R)}^{(K)} \leq \sum_{r \in m} C_{(r)}^{(1)} \leq \sum_{r \in m} \widetilde{C}_{(r)}^{(1)}. \quad (3)$$

Applying Eq. 2 to $m = \widehat{m}_{t_1,t_2}^{(K)}$ and then Eq. 3 to $m = \widetilde{m}_{t_1,t_2}^{(K)}$ we get the theorem. ∎

## 3  Joint Segmentation and Parallelization of the Algorithm

Joint segmentation arises when more than one profile should be segmented jointly. We make the distinction between simultaneous segmentation, where all breakpoints are the sames across profiles, with joint segmentation where all profiles have their own specific breakpoints, but may share some characteristics, like the same noise, the same biases, the same values for the mean of segments that share the same copy numbers (i.e. parameters $\mu_p$). When segmenting $I$ profiles jointly, a typical model is:

$$\forall i \in [1, I], \; \forall t \in r_k^i, \; Y_i(t) | \{Z(r_k^i) = p\} \sim \mathcal{N}(\mu_p + b(t), \sigma^2),$$

where $r_k^i$ stands for segment $k$ of profile $i$, and where $b(t)$ is a bias function that depends on the position (like the wave effect [9]). Then the segmentation of profile $i$ into $K_i$ segments is denoted by $m_i^{K_i} = \{r_1^i, ..., r_{K_i}^i\}$, and the *global* segmentation into $K$ segments is denoted by $m_K = \{m_1^{K_1}, ..., m_I^{K_I}\}$, with $K = \sum_{i=1}^{I} K_i$.

Joint segmentation presents an additional algorithmic challenge. When each $K_i$ is known, the best segmentation for each profile $\widehat{m}_i^{K_i}$ can be found independently and therefore computed in parallel using Algorithm 1 or 2. Then the additional step, which remains sequential, is to determine (i) the common parameters across profiles ($\{\mu_p\}, b(t)$) and (ii) the best combination of $\{\widehat{K}_i\}$ that provides the best joint segmentation for a given total number of segments $K$ [9]. Dynamic Programming has also been shown to provide an exact solution to problem (ii) in $\mathcal{O}(I^3 K_{\max}^2)$, where $K_{\max}$ is the maximum number of segments to be put in each profile. Consequently, DP-based joint segmentation alternates between parallel steps (computation of individual segmentations) and sequential steps (estimation of common parameters and determination of the best combination of individual segmentations), as shown in Algorithm 3.

---

**Algorithm 3.** Parallel Algorithm for Joint segmentation

---

**Input:** $\{Y_i(t)\}$, $I$ profiles of $n$ observations, $K$ an integer, $\{\widehat{\mu}_p^0\}$ starting values for common parameters
**Input:** $C_{t_1, t_2}^i$ cost of the segments $]t_1, t_2]$ for profile $i$ for all $(t_1, t_2) \in [\![1, n]\!]^2$
**Output:** $\{\widehat{K}_1, ..., \widehat{K}_I\}$, $C_{1,t}^{(\widehat{K}_i)}$ in $\mathbb{R}$ for all $i \leq I$, $k \leq K_i$ and $t \leq n$, $\{\widehat{\mu}_p\}$, $\widehat{b}(t)$
**while** not convergence **do**
    **for** $i \in \{1, ...I\}$ **do in parallel**
        compute $C_{1,t}^{(k)}$ $\forall k \leq K$ and $t \leq n$ with Algorithm 2
    **end for**
    update $\{\widehat{\mu}_p\}, \widehat{b}(t)$
    compute $\{\widehat{K}_1, ..., \widehat{K}_I\}$ with **sequential** DP [9]
**end while**

---

# 4   Correctness, Computational Footprint and Scalability

All the presented algorithms are available into the `cghseg` R-package[1], which now relies on the **parallel** R-package and on the shared memory programming

---

[1] http://cran.r-project.org/web/packages/cghseg

**Table 1.** Performance comparison between the non linearized and non parallel ("old") and the linearized-parallel ("new") versions. Performance is assessed through Mean Square Errors (over 50 replicates) for the estimates of the number of segments ($\hat{K}$), the mean level of each segment ($\hat{\mu}$), and confidence intervals for the False Discovery and False Negative Rates for breakpoint detection. The methodology is similar to [9].

| SNR | $MSE_{new}(\hat{K})$ | $MSE_{old}(\hat{K})$ | $MSE_{new}(\hat{\mu})$ | $MSE_{old}(\hat{\mu})$ | $CI_{new}(fdr)$ | $CI_{old}(fdr)$ | $CI_{new}(fnr)$ | $CI_{old}(fnr)$ |
|---|---|---|---|---|---|---|---|---|
| 1 | [0.90; 1.02] | [0.90; 1.02] | [0.14; 0.16] | [0.14; 0.17] | [0.42; 0.45] | [0.42; 0.45] | [0.59; 0.62] | [0.59; 0.62] |
| 5 | [0.36; 0.44] | [0.37; 0.45] | [0.05; 0.06] | [0.05; 0.06] | [0.15; 0.20] | [0.17; 0.22] | [0.21; 0.26] | [0.23; 0.28] |
| 10 | [0.22; 0.30] | [0.23; 0.30] | [0.03; 0.03] | [0.03; 0.03] | [0.06; 0.10] | [0.07; 0.11] | [0.08; 0.14] | [0.09; 0.14] |
| 15 | [0.17; 0.22] | [0.17; 0.23] | [0.02; 0.02] | [0.02; 0.02] | [0.02; 0.06] | [0.03; 0.07] | [0.03; 0.08] | [0.03; 0.09] |
| 20 | [0.15; 0.20] | [0.15; 0.21] | [0.01; 0.02] | [0.01; 0.02] | [0.01; 0.05] | [0.01; 0.05] | [0.01; 0.06] | [0.01; 0.07] |

standard openMP in the C++ sections. The new package is now designed to be executed in parallel on multiprocessor architectures.

We recall that the statistical performance of the model have been discussed elsewhere [9], so that our focus here is computational only. Consequently, we use a previously published simulation scheme to generate the data [9,12] with $n = 20,000$ observations per profile, and a number of profiles of 256, 512 and 1024. The average number of segments is set to 10 for each profile, and the Signal to Noise Ratio is set to 5 (which corresponds to moderately easy configurations [9]).

We first check the correctness of our method, by verifying that the linearization approximations (with and without calling) give the same statistical performance compared with the non-linearized version. This is shown in Table 1), as the performance are identical over 50 replicates.

Then we assess the effectiveness of the parallel implementation by running cghseg on an increasing number of cores (1, 2, 4, 16, 32, 48). When dealing with parallelized codes, the Amdahl's law [1] provides the expected theoretical speedup with respect to the time proportion of the sequential part and the number of cores. The speedup of cghseg follows the Amdahl's law when the number of profiles is high (Fig. 1, dashed lines), which demonstrates the quality of our implementation as the Amdahl's law constitutes the best possible speedup. However, we observe an unexpected moderate speedup decrease for a lower number of profiles. This is due to overheads associated with the use of the parallel R-package, overheads which become negligible when the number of profiles is high. Still, the execution time of configurations with 256 profiles is 6 min on average using 48 cores, which remains excellent.

While the main interest of cghseg lies in the quality of its results, the associated computational expense is affordable even for very large datasets. As a benchmark, we simulated datasets with 1024 profiles of length 100,000 which corresponds to the up-to-date limits of the available datasets of SNP-arrays for instance. When joint segmentation is performed along with calling, cghseg required 4 h on average on 48 cores (see Table 2), which is very reasonable considering the size of the dataset. Lastly, due to the *copy-on-write* mechanism of the parallel R-package that avoids memory copy between processes, and thanks to the shared memory efficiency provided by openMP, the memory needs of cghseg

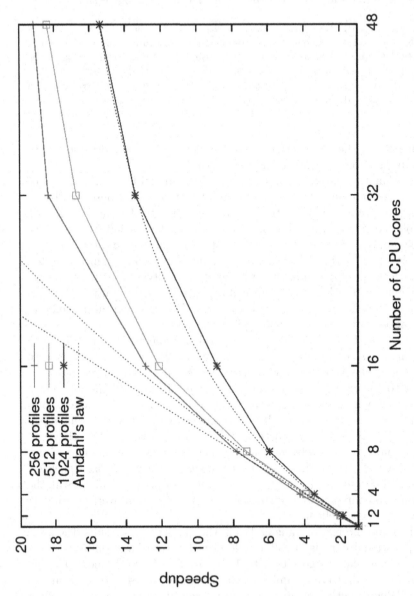

**Fig. 1.** Average speedup of cghseg observed on simulated datasets for a varying number of profiles and available cores (plain lines). Amdahl's law is empirically adjusted to the computing results with 256, 512 and 1024 profiles (dashed lines, with $\alpha = 0.2, 1.2, 4.5\%$ respectively). The theoretical speedup is given by $1/(\alpha + (1 - \alpha)/c)$, where $\alpha$ is the time proportion of the sequential part and $c$ the number of cores. Points correspond to averages over 5 replicates. Run on a quadri-12 cores Opteron 2.2 GHz, 256 Gb RAM.

**Table 2.** Average computational requirements of `cghseg` estimated on simulated datasets for a varying number of profiles and observations, computed on 48 cores. Results correspond to averages over 5 replicates. Run on a quadri-12 cores Opteron 2.2 GHz, 256 Gb RAM.

| $n$ (observations/profile) | 20,000 | | | 100,000 | | |
|---|---|---|---|---|---|---|
| $I$ (number of profiles) | 256 | 512 | 1024 | 256 | 512 | 1024 |
| Average CPU time (min) | 6 | 15 | 54 | 31 | 70 | 253 |
| Memory usage (Gb) | 0.4 | 0.8 | 1.8 | 1.7 | 3.7 | 7.9 |

only correspond to the dataset under study (see Table 2). Therefore, our method and software package are adapted to the next generation of computers (many cores) and experiments (large profiles).

# References

1. Amdahl, G. M.: Validity of the single processor approach to achieving large scale computing capabilities. In: Proceedings of the AFIPS '67 Spring Joint Computer Conference, 18–20 April 1967 (Spring), pp. 483–485. ACM (1967)
2. David, L., Huber, W., Granovskaia, M., Toedling, J., Palm, C.J., Bofkin, L., Jones, T., Davis, R.W., Steinmetz, L.M.: A high-resolution map of transcription in the yeast genome. Proc. Natl. Acad. Sci. USA **103**(14), 5320–5325 (2006)
3. Hocking, T.D., Schleiermacher, G., Janoueix-Lerosey, I., Delattre, O., Bach, F., Vert, J.-P.: Learning smoothing models using breakpoint annotations. HAL Technical report 00663790 (2012)
4. Killick, R., Fearnhead, P., Eckley, I. A.: Optimal detection of changepoints with a linear computational cost. arXiv:1101.1438, January 2011.
5. Marioni, J.-C., Thorne, N.-P., Tavare, S.: BioHMM: a heterogeneous hidden markov model for segmenting array CGH data. Bioinformatics **22**(9), 1144–1146 (2006)
6. Nicolas, P., Bize, L., Muri, F., Hoebeke, M., Rodolphe, F., Ehrlich, S.D., Prum, B., Bessieres, P.: Mining Bacillus subtilis chromosome heterogeneities using hidden Markov models. Nucleic Acids Res. **30**(6), 1418–1426 (2002)
7. Nicolas, P., Leduc, A., Robin, S., Rasmussen, S., Jarmer, H., Bessieres, P.: Transcriptional landscape estimation from tiling array data using a model of signal shift and drift. Bioinformatics **25**(18), 2341–2347 (2009)
8. Olshen, A.B., Venkatraman, E.S., Lucito, R., Wigler, M.: Circular binary segmentation for the analysis of array-based DNA copy number data. Biostatistics **5**(4), 557–572 (2004)
9. Picard, F., Lebarbier, E., Hoebeke, M., Rigaill, G., Thiam, B., Robin, S.: Joint segmentation, calling and normalization of multiple array CGH profiles. Biostatistics **12**(3), 413–428 (2011)
10. Picard, F., Robin, S., Lavielle, M., Vaisse, C., Daudin, J.-J.: A statistical approach for array CGH data analysis. BMC Bioinf. **6**, 27 (2005)
11. Picard, F., Robin, S., Lebarbier, E., Daudin, J.-J.: A segmentation/clustering model for the analysis of array CGH data. Biometrics **63**, 758–766 (2007)

12. Pique-Regi, R., Ortega, A., Asgharzadeh, S.: Joint estimation of copy number variation and reference intensities on multiple DNA arrays using GADA. Bioinformatics **25**(10), 1223–1230 (2009)
13. Rigaill, G.: Pruned dynamic programming for optimal multiple change-point detection. arxiv:1004.0887, April 2010
14. Shah, S.P.: Computational methods for identification of recurrent copy number alteration patterns by array CGH. Cytogenet. Genome Res. **123**(1–4), 343–351 (2008)
15. Teo, S.M., Pawitan, Y., Kumar, V., Thalamuthu, A., Seielstad, M., Chia, K.S., Salim, A.: Multi-platform segmentation for joint detection of copy number variants. Bioinformatics **27**(11), 1555–1561 (2011)
16. van de Wiel, M.A., Picard, F., van Wieringen, W.N., Ylstra, B.: Preprocessing and downstream analysis of microarray DNA copy number profiles. Brief. Bioinf. **12**(1), 10–21 (2011)
17. van de Wiel, M.A., Kim, K.I., Vosse, S.J., van Wieringen, W.N., Wilting, S.M., Ylstra, B.: CGHcall: calling aberrations for array cgh tumor profiles. Bioinformatics **23**(7), 892–894 (2007)
18. Willenbrock, H., Fridlyand, J.: A comparison study: applying segmentation to array CGH data for downstream analyses. Bioinformatics **21**(22), 4084–4091 (2005)
19. Zhang, N.R., Siegmund, D.O.: A modified Bayes information criterion with applications to the analysis of comparative genomic hybridization data. Biometrics **63**(1), 22–32 (2007)
20. Zhang, N.R., Siegmund, D.O., Ji, H., Li, J.Z.: Detecting simultaneous changepoints in multiple sequences. Biometrika **97**(3), 631–645 (2010)

# Identification of Pathway Signatures in Parkinson's Disease with Gene Ontology and Sparse Regularization

Margherita Squillario[✉], Grzegorz Zycinski[✉], Annalisa Barla, and Alessandro Verri

DIBRIS, University of Genoa, Via Dodecaneso 35, 16146 Genoa, Italy
{margherita.squillario,grzegorz.zycinski,
annalisa.barla,alessandro.verri}@unige.it
http://www.dibris.unige.it/

**Abstract.** The purpose of this work is to compare Knowledge Driven Variable Selection (KDVS), a novel method for biomarkers, processes and functions identification with the most frequently used pipeline in the analysis of high–throughput data (Standard pipeline). While in the Standard pipeline the biological knowledge is used after the variable selection and classification phase, in KDVS it is used a priori to structure the data matrix. We analyze the same gene expression dataset using $\ell_1\ell_2^{FS}$, a regularization method for variable selection and classification, choosing Gene Ontology (GO) as source of biological knowledge. We compare the lists identified by the pipelines with state–of–the–art benchmark lists of genes and GO terms known to be related with Parkinson's disease (PD). The results indicate that KDVS performs significantly better than the Standard pipeline.

**Keywords:** Knowledge driven variable selection · $\ell_1\ell_2^{FS}$ · Enrichment · Benchmark · Parkinson · Gene Ontology · Biomarkers

## 1 Introduction

Parkinson's disease (PD) is a neurodegenerative disorder that impairs the motor skills at the onset and successively the cognitive and the speech functions. Like other neurological diseases [1,2], once the first symptoms appear, a great loss of neurons has already occurred and so far no clinical tests are able to early diagnose it. In this scenario, it is relevant to identify those altered processes, functions for which candidate biomarkers can be discovered, especially in the early onset of PD.

To this aim, we analyze an early onset gene expression microarray dataset, containing cases (i.e., diseased) and controls (i.e., healthy) samples, comparing two different approaches. We refer to the most frequently used approach as Standard, and we present here an alternative method, Knowledge Driven Variable Selection (KDVS) [3,4] (see Fig. 1).

© Springer International Publishing Switzerland 2014
E. Formenti et al. (Eds.): CIBB 2013, LNBI 8452, pp. 259–273, 2014.
DOI: 10.1007/978-3-319-09042-9_19

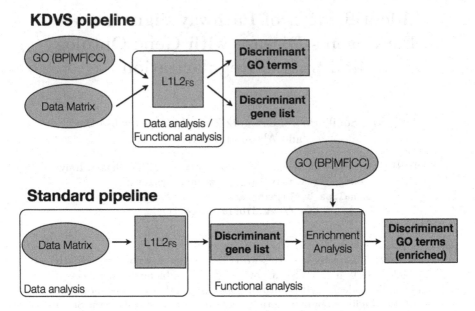

**Fig. 1.** KDVS and Standard pipelines. Gene Ontology (GO) is divided in three domains: Biological Process (BP), Molecular Function (MF) and Cellular Component (CC).

Some steps are common for both pipelines. First, the gene expression data are collected from diseased patients and healthy controls. Next, the data are normalized to produce uniform gene expression data matrix, where columns refer to biological samples, and rows refer to genes measured.

In the Standard pipeline, classification and subsequent variable selection [5] (i.e., data analysis) are performed on gene expression data matrix, producing a list of selected variables (i.e., genes), that maximally discriminates two physiological states (i.e., cases and controls). Finally, the list of selected genes (i.e., gene signature) is enriched [6] (i.e., functional analysis) using a chosen source of biological knowledge to determine the functions or processes, in which the discriminant genes are involved in. The final result consists of a list of genes, that are probably associated to the development of the investigated biological state, and a list of enriched functions, processes or interactions, characterizing the disease, that depend on the source of biological knowledge utilized. In this work, the enrichment analysis for the Standard pipeline was performed using the webtoolkit WebGestalt [7], with GO as source of domain knowledge.

In general, the two most used sources of biological knowledge are Gene Ontology (GO) [8] and the Kyoto Encyclopedia of Genes and Genomes (KEGG) [9]. We choose GO as source of biological knowledge for both pipelines, considering all three domains that constitute GO: Molecular Functions (MF), Biological Process (BP) and Cellular Component (CC).

The novel alternative method, KDVS, uses a priori biological knowledge, before the classification and variable selection phase, rather than a posteriori in the enrichment step, as in the Standard pipeline. Briefly, the genes are associated to GO terms through a curated process of annotation. For each GO term in a given domain, individual submatrices of the gene expression data matrix are produced. Each of these submatrices, containing all the samples and all the genes annotated to a specific GO term, is then the subject of variable selection and classification. The partial results from all submatrices are integrated into a final result. Each partial result consists of a classification error and a list of selected variables (i.e., genes) that contribute most to the discrimination between two classes. The final result consists of a list of GO terms that are identified as discriminating between the two classes, and of a gene signature, derived from the combination of all the individual gene lists produced for each selected GO term.

The variable selection method of choice in both pipelines is $\ell_1\ell_{2FS}$ [10], a regularization method that belongs to the class of embedded methods [5], which incorporate variable selection within the classification step. We choose this method because many state–of–the–art methods (e.g., t-test, Wilcoxon, ANOVA) do not cope well with multivariate models, which are more suited to reflect the behavior of genes involved in a complex disease (e.g., tumors, neurodegenerative diseases). The major drawback of univariate models, often used as standard approaches in high-throughput data analysis, concerns their inability to exploit the complex correlation among molecular variables, especially in the context of multifactorial diseases. Also, recent publications [11,12] proved the suitability of multivariate models in identifying reliable gene signatures.

In this paper, we compare the results of KDVS and Standard pipeline in the analysis of the same gene expression data set, investigating which pipeline provides as output more biologically sound results. Both pipelines identify a list of discriminant GO terms and a list of selected genes, one using the available biological knowledge *a priori* and the other one *a posteriori*. To validate these findings *in silico*, we compare the results obtained from each pipeline with the currently available biological knowledge regarding PD. To this aim, we compile two *benchmark lists*: one of GO terms, and one of genes, relevant to PD according to the current state of research.

Our experiments show that KDVS, the new proposed pipeline, identifies more discriminating GO terms and genes, known to be related to PD, with respect to the Standard pipeline. This result is confirmed by the precision, recall and F-measures calculated on the results derived from the pipelines.

## 2    Materials and Methods

For our experiments, we use a public microarray data set from the Gene Expression Omnibus (GEO) [13]. The dataset (ID: GSE6613, [14]) is composed of measurements on whole blood samples from 50 patients affected by early onset PD and from 22 healthy controls. The microarray platform is the Affymetrix Human

Genome U133A Array (HG-U133A) and contains 22283 probesets, that measure the gene expression. We devise a binary classification problem of healthy vs. diseased samples. Hence, our data matrix is $72 \times 22283$-dimensional. Normalization of the gene expression values is performed using the Robust Multichip Average method [15], with a R script based on the *aroma* package[1].

## 2.1    Feature Selection Framework

$\ell_1\ell_{2FS}$ is an embedded regularization method for variable selection capable to identify subsets of discriminative genes. It can be tuned to give a minimal set of discriminative genes or larger sets including correlated genes [16]. The method is based on [17] and it was successfully applied in the analysis of molecular high–throughput data [11,12].

The objective function is a linear model $f(x) = \beta x$, whose sign gives the classification rule that can be used to associate a new sample to one of the two classes. The sparse weight vector $\beta$ is found by minimizing the $\ell_1\ell_2$ functional:

$$||Y - \beta X||_2^2 + \tau||\beta||_1 + \mu||\beta||_2^2$$

where the least square error is penalized with the $\ell_1$ and $\ell_2$ norm of the coefficient vector $\beta$. Once the relevant features are selected we use regularized least squares (RLS) to estimate the classifier:

$$||Y - \tilde{X}\beta||_2^2 + \lambda||\beta||_2^2$$

where $\tilde{X}$ is the submatrix obtained by only using the columns of $X$ corresponding to the variables selected at the first step. The training for selection and classification requires a careful choice of the regularization parameters $\tau$ and $\lambda$ for both $\ell_1\ell_2$ and RLS. The third parameter $\mu$ is fixed a priori and governs the amount of correlation allowed for selected variables. To avoid biased results [18], model selection and statistical significance assessment are performed within two nested $K$-cross validation loops as in [10]. The framework is implemented in Python and uses the L1L2Signature[2] and L1L2Py[3] libraries.

## 2.2    The KDVS Pipeline

The general schema of KDVS [4] is based on the prototype presented in [3] and is depicted in Fig. 2. KDVS consists of a *local integration framework, KDVS core* that performs an integration of microarray platform data and annotations [13] with prior biological knowledge. Outside this framework, the raw data are pre-processed for normalization and summary with state–of–the–art algorithms for microarray technologies [19]. *KDVS core* performs all data management operations and provides skeletal execution environment for user applications. The data

---

[1] http://www.aroma-project.org
[2] http://slipguru.disi.unige.it/Software/L1L2Signature
[3] http://slipguru.disi.unige.it/Software/L1L2Py

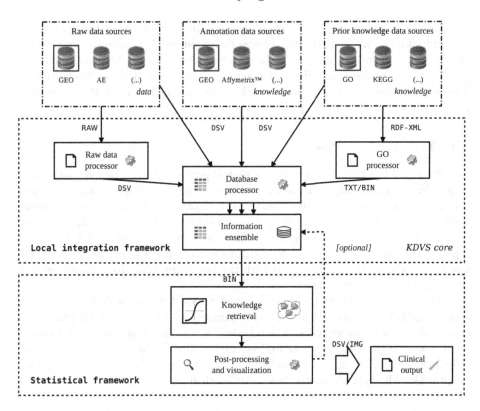

**Fig. 2.** Schema representing the KDVS inner structure.

management part consists of several processors that work with specific type of data. It also maintains the information ensemble, a shared data storage that can be used across user applications. *KDVS core* provides convenient functionalities for developers of user applications, such as atomic execution units, variable sharing, interface to parallel computational resources etc. The result of the local integration framework consists in extracting a submatrix: taking for each GO term, the corresponding set of probesets and the expression values extracted across all samples. The submatrix is fed as input of the classification/feature selection task performed by the *Statistical framework*, where the knowledge discovery concept is implemented.

The *Statistical framework* runs the variable selection/classification method of choice on each submatrix, in our case $\ell_1\ell_2{}_{FS}$. Each subproblem is processed and treated independently as a variable selection and classification problem. Partial results (gene signatures, classification errors, performance plots) are collected and postprocessed to obtain the final output.

KDVS[4] is implemented in Python and runs in parallel in order to speed–up the computation. In the current implementation, an *ad–hoc* environment is built over a local network of multicore desktop machines, where the computational tasks were distributed to individual machines and executed in the background. The environment is controlled using the Python package PPlus[5], a Python wrapper for the python library *Parallel Python*.

## 2.3 The Standard Pipeline

The Standard pipeline represents the classical procedure to extract relevant biological features from high–throughput datasets and it is composed of two steps: the *data analysis* and the *functional analysis* [20]. In the first one, the data matrix is normalized and then analyzed with $\ell_1\ell_2{}_{FS}$. The typical output of this phase is a list of discriminant genes, defined as gene signature, that is given as input to the *functional analysis*, whose purpose is to find the statistically significant categories that are associated with the gene signature. To this aim we use the online toolkit WebGestalt[6].

This web-service takes as input a list of relevant genes/probesets and performs a GSEA analysis [21] in several databases, as KEGG or GO, identifying the most relevant pathways and ontologies in the signatures. For our experiments, we use the GO database and select the WebGestalt human genome as reference set, 0.05 as level of significance, 3 as the minimum number of genes and the default Hypergeometric test as statistical method.

## 2.4 Benchmark Lists

To obtain the necessary benchmark lists we followed the workflow in Fig. 3. We started with the union (in terms of sets) of three gene lists: the first one from the "Parkinson's disease - Homo sapiens" pathway of KEGG PATHWAY database (ID: hsa05012), the second one from the "Parkinson's disease (PD)" entry of KEGG DISEASE database (ID: H00057), and the third one from the result of Gene Prospector tool [22] when queried for "Parkinson's disease". In case of KEGG databases, the genes associated through "linkDB" were included in the respective lists. While the first two lists contain genes that have been experimentally verified to be involved in the disease, the third list could contain also genes that are loosely connected to the disease, because they are derived from high–throughput experiments and therefore need further experimental validation. Next, we merged the three lists and eliminated duplicates. The merged list contained a total of 482 genes.

To proceed, we used Gene Ontology Annotations (GOA), compiled for *homo sapiens*. Here, each gene is associated with some GO term(s), based on specific

---

[4] http://www.scfbm.org/content/supplementary/1751-0473-8-2-s1.zip
[5] http://slipguru.disi.unige.it/Software/PPlus
[6] http://bioinfo.vanderbilt.edu/webgestalt/

evidences. The evidences describe the work or analysis upon which the association between a specific GO term and gene product is based; there are 22 possible evidence codes in GO[7]. Each single association of a gene to an evidence is tagged with the annotation date as well.

Based on that information, we constructed a filtering schema to derive the *benchmark gene list* and *benchmark GO terms list*, that are strongly associated with PD. The first benchmark list is composed of 444 genes and the second of 2121 GO terms. Two filters were applied: based on the annotation date (for GO terms), and based on the evidence strength (for both genes and GO terms).

The first filter was applied because, while examining the annotations, we notice that each gene can be associated with the same GO term but with different evidences, due to the internal history of GO curation process. During the construction of the benchmark lists, we kept those associations whose evidences displayed the most recent annotation date.

The second filter was applied based on the reliability of the evidences [8]. Based on this consideration, we arbitrarily defined the trustability of the evidences, as follows: the evidences recognized as more trustable include all those belonging to the Experimental Evidence Codes' category (i.e., EXP, IDA, IPI, IMP, IGI, IEP), the Traceable Author Statement (i.e., TAS), and the Inferred by Curator (i.e., IC). During the construction of benchmark lists, we kept the genes and GO terms associated with this selection of evidences.

## 2.5   Precision, Recall and F-measure

Precision and recall are evaluated following their usual definition:

$$\text{Precision} = tp/(tp + fp)$$

$$\text{Recall} = tp/(tp + fn)$$

We consider as *true positives* the genes or GO terms that were selected as relevant and that belong to the corresponding benchmark list. *False positives* genes or GO terms are those selected items that do not belong to the corresponding benchmark list. *False negatives* genes or GO terms are those items of the corresponding benchmark list that were not selected by the pipeline.

For example, in case of the Standard pipeline, the precision for the genes was 3/66 and for the GO terms was 31/65. The recall for the genes was 3/444, while for the GO terms was 31/2121 (Table 1).

For KDVS, we compute precision and recall for the cumulative list of genes and GO terms (Table 1) and for each domain separately (Table 2).

The F-measure follows its definition:

$$\text{F-measure} = 2 \times \text{Precision} \times \text{Recall}/(\text{Precision} + \text{Recall}).$$

---

[7] ftp://ftp.geneontology.org/go/www/GO.gettingStarted.shtml

# 3   Results and Discussion

## 3.1   The KDVS Pipeline

The analysis of the early onset PD microarray data set with KDVS provided a list of discriminant terms for each domain of GO: 150 for MF, 418 for BP and 103 for CC. A term was called *discriminant*, if the classification performed on the corresponding data submatrix resulted in an accuracy of at least 70 %, i.e. an error below 30 % (for this dataset, the equal error rate was 55 %). To ensure unbiased results, for each variable selection problem we used a nested double cross validation procedure with 9 external and 8 internal splits.

From the list of discriminant GO terms, a list of selected probesets (and related genes) was derived: 6072 probesets, associated to 4286 genes for MF, 3822 probesets, associated to 2713 genes for BP and 3842 probesets, associated to 2836 genes for CC.

## 3.2   Comparison Between Lists: KDVS and Benchmark

To validate the biological soundness of the GO term lists identified by KDVS, we evaluated its intersection with the benchmark list of GO terms compiled for PD (*term benchmarking*), following the workflow depicted in Fig. 3 (see Materials and Methods).

Since the KDVS pipeline was run separately for MF, BP and CC domains, the comparison between the two lists was restricted only to the elements belonging to

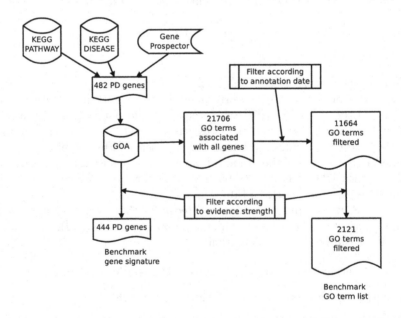

**Fig. 3.** Workflow followed to build the *benchmark* gene and GO terms lists.

each specific domain knowledge. When MF was considered, the overlap consisted of 54 terms, corresponding to a 12 % coverage of the benchmark list. Here, the terms were related to the binding to motor proteins (i.e., kinesin and dynein), to components of the extracellular matrix and of the cytoskeleton (respectively fibronectin and beta–tubulin), to binding ions and groups (i.e., iron, manganese, magnesium, heme), to binding to nucleotidic acids (i.e., chromatin, rRNA and the mRNA 3′ UTR), and finally to binding to proteins like chaperone, ubiquitin and syntaxin. The activities above involved mainly enzymes (e.g., hydrolase, peptidase, especially serine–type peptidase, peroxidase), but also transcription factors, such as ubiquinone and ubiquitin.

When BP was considered, the overlap consisted of 154 terms, correspond-ing to a 11 % coverage of the benchmark list. Here, the terms were related to various kind of metabolic processes concerning lipids (e.g., prostaglandins, sphin-golipids), ATP or dopamine; to the development of the brain, specifically to the neuromuscular junctions, and the cardiac muscle tissue; to defense response, in particular from viruses, that prompt the activation and regulation of B and T cells. The biological terms related to the cell are those that control cell death, arrest, growth, differentiation, proliferation and specific cell cycle checkpoints (e.g., from G1 to S). The involved pathways concerned: the Notch receptors, which regulate cell-cell communication in several ways (acting in particular in the central nervous system and in the heart); the Wnt proteins, which regulates cell–cell communication either, passing the information from the plasma mem-brane to the nucleus, prompting the transcription regulation of specific genes; three specific groups of growth factors, which are nerve, platelet–derived and insulin–like, and are fundamental respectively for the growth, maintenance, and survival of neurons, for the angiogenesis or the growth of existing blood vessels, and for the communication with the environment, through the promotion of cell proliferation and inhibition of cell death.

When CC was considered, the overlap consisted of 34 terms, corresponding to a 15 % coverage of the benchmark list. The terms were particularly related to the mitochondrion (e.g., matrix cellular, outer membrane cellular, proton–transporting ATP synthase complex cellular component), to the cell–cell bound-aries (e.g., focal adhesion, basal lamina), to the vesicles and the methods of transport of molecules inside the cell (e.g., coated pit, vesicle, clathrin–coated vesicle).

We also evaluated the overlap between the lists of discriminant genes and the benchmark list of genes, compiled for PD (*gene benchmarking*), following the custom workflow in Fig. 3 (see Materials and Methods). Considering MF as source of biological knowledge, the overlap consisted of 196 genes, corresponding to 44 % of the benchmark list. Using BP, the overlap consisted of 161 genes, corresponding to 36 % of the benchmark list. With CC, the overlap consisted of 123 genes, corresponding to 28 % of the benchmark list.

We calculated the precision, recall and F-measures as shown in Materials and methods, considering, for KDVS, MF, BP, CC together (Table 1) and separately (Table 2). These values were calculated to evaluate the classification performance

**Table 1.** Comparison of precision, recall and F-measures for both pipelines, considering all the GO terms.

|  | Standard | KDVS |
|---|---|---|
| Precision GO terms | 47.7 % | 36 % |
| Recall GO terms | 1 % | 11.4 % |
| Precision genes | 4.5 % | 7.8 % |
| Recall genes | 0.7 % | 16.2 % |
| F-measure GO terms | 0.028 | 0.173 |
| F-measure genes | 0 | 0.108 |

**Table 2.** Comparison of precision, recall and F-measures for KDVS, considering the three domains of GO separately.

|  | KDVS | | |
|---|---|---|---|
| GO domain | MF | BP | CC |
| Precision GO terms | 36 % | 36.8 % | 32.7 % |
| Recall GO terms | 2.5 % | 7.3 % | 1.6 % |
| Precision genes | 4.6 % | 5.9 % | 4.3 % |
| Recall genes | 44.1 % | 36.3 % | 27.7 % |
| F-measure GO terms | 0.047 | 0.122 | 0.029 |
| F-measure genes | 0.082 | 0.1 | 0.075 |

of both pipelines, considering the state–of–the–art knowledge on PD, here represented by the benchmark lists.

If we compare the results in Table 2 between GO terms and genes, it is evident that the precision is higher than the recall in the GO terms case, while an opposite trend is evident in the genes case. This observation could be argued considering the known redundancy of the genes: the low precision obtained for the genes, could be explained by the fact that the same role can be performed by more than one gene. Therefore the processes or functions characterizing a specific disease constitute a well defined group, but there is a high number of possible and different combinations of genes that can perform one specific function or that can be involved in a specific process.

## 3.3   The Standard Pipeline

## 3.4   Data Analysis

The analysis of the early onset PD microarray data set with the Standard pipeline provided a signature of 77 probesets associated to 66 genes. To ensure unbiased results, for the variable selection step we used a nested double cross validation procedure with 9 external and 8 internal splits, resulting in a 64 % accuracy performance.

## 3.5    Functional Analysis

The functional analysis was performed with WebGestalt [7], using GO as source of biological knowledge; 65 terms were identified as enriched. The terms from the MF domain showed that the majority of the proteins are involved in the binding to other molecules (e.g., proteins or nucleic acids) or to metals. This last function has a fundamental role in generating free radicals responsible of oxidative stress and was already shown to be important in PD and in other neurodegenerative diseases [23]. Furthermore the generation of protein aggregates, shown from some discriminating MF terms (e.g., protein homodimerization activity, identical protein binding) is also a known feature of PD [24]. The terms from the CC domain showed that most of the identified gene products are located in the cytoplasm, and are also associated with the cytoskeleton, whose involvement (in particular the microtubules) is becoming an emerging topic in the pathogenesis of PD [25]. The terms from the BP domain concerned the response to different stimuli (i.e., *stress, chemical and other organisms*), the involvement of the immune system (both in its development and in the defense response), the cell death, the biological regulation of homeostatic processes (in particular of the erythrocytes), and of metabolism. Both the terms regarding responses to *chemical stimulus* and *other organism* (in our case *virus*), are the two candidate processes suggested to be external triggers for initiating the pathogenesis of PD [26].

## 3.6    Comparison Between the Lists: Standard and Benchmark

To assess how much of the GO terms identified by the Standard pipeline were known to be involved in PD, we performed an overlap between its list and the benchmark list of GO terms compiled for PD (*term benchmarking*), following the workflow depicted in Fig. 3. The overlap between the lists was of 31 GO terms. The terms from the BP domain were mainly related to the defense response in general and to the response to virus, to the activation of the immune system, particularly the innate part, and to the regulation of the apoptosis. The terms from the MF domain were related to the binding to other proteins (e.g., actin, kinase), to binding to iron and heme group, and to structural component of the cytoskeleton. The terms from the CC domain were related to the cytoskeleton, the cell cortex, the basolateral plasma membrane and the lysosome.

Similarly, we evaluated the intersection between the gene list, identified by WebGestalt, and the benchmark list of genes (*gene benchmarking*). The identified lists had three common genes, namely SNCA, ATXN1 and HLA-DQB1.

The values shown in Table 1 underline the fact that the precision calculated on the GO terms identified was sixteen times higher than the precision computed considering the genes. One possible explanation can be attribute to the gene redundancy, already mentioned above.

## 3.7    Comparison of KDVS and Standard Pipelines

The final goal of our analysis was to evaluate how well KDVS captures the known biological knowledge about PD, in comparison with the Standard approach.

To this aim, we compared the lists of the pipelines with the benchmark lists, that we defined following the workflow in Fig. 3 (see Materials and Methods).

This approach provided a way to validate the results *in silico*, using state-of-the-art information about the genes, the functions and the processes known to be involved in the disease. We also calculated the precision, recall and the F-measure to accurately validate our results from both the statistical and the biological viewpoints (Tables 1, 2).

The comparisons showed that the results obtained with KDVS have a sensibly higher information overlap with the trustable knowledge gathered for PD (here represented by the benchmark lists), than the results of the Standard pipeline. It is evident that, considering the GO terms, the precision was higher in the Standard pipeline than in KDVS but the recall was higher in KDVS. This underlines that KDVS returns more relevant results (considering the recall) with respect to the other pipeline, but with a slightly lower precision. For the genes, both precision and recall were considerably lower with respect to the results obtained with the GO terms, but they were higher for KDVS than for the Standard pipeline. Again, the recall was much higher for KDVS than for the other approach. Looking at this data it is clear that the Standard pipeline identified fewer genes in comparison with KDVS but involved in very few and highly specific GO terms. The KDVS pipeline, instead, identified more genes with respect to Standard, but involved in a relevant number of GO terms, less specific compared to the other approach.

Finally, we performed a three–way comparison between the lists of GO terms from Standard, KDVS and benchmark lists. This revealed thirteen common terms. The most relevant MF terms were: *iron ion binding, heme binding, identical protein binding, protein homodimerization activity.* The most relevant BP terms were: *cell death, response to stress, defense response* and *defense response to virus.*

The same kind of three–way comparison was performed, considering the respective lists of genes. Considering MF as source of domain knowledge, this comparison gave SNCA and ATXN1 as common genes. When using BP just ATXN1 was found in common among the lists, using CC, ATXN1, SNCA and HLA-DQB1 were in common. While SNCA is a gene already known to be associated with PD [27], ATXN1 is prevalently known to be associated with another neurodegenerative disorder, known as Autosomal Dominant Cerebellar Ataxia (ADCA) [28]. The main clinical symptom of this disease is ataxia, that consists of gross lack of muscle movements coordination that is severely impaired also in PD patients.

## 4   Conclusions

The purpose of this work was to identify possible candidate biomarkers, processes and functions related with PD by comparing the results of the new proposed pipeline, namely KDVS, with the Standard pipeline, usually used to analyze high–throughput data.

The KDVS pipeline represents a novel approach to analyze gene expression data because, differently from the Standard pipeline, combines in one single step the functional and the data analysis (see Fig. 1). In KDVS, the data matrix is used as a template to generate data submatrices that correspond to expression of genes related to individual GO terms. Each data submatrix, containing the same number of samples, is analyzed by the $\ell_1\ell_{2FS}$ regularization method to identify discriminant GO terms, as well as the relevant genes associated with those terms, defined as gene signature.

Conversely, in the Standard pipeline, the data matrix is first analyzed by $\ell_1\ell_{2FS}$ to obtain the gene signature, then the signature is functionally characterized by an enrichment procedure in order to identify processes and functions in which the discriminant genes are involved in. Such characterization was performed with the public on–line tool WebGestalt [7], that utilizes the gene set enrichment analysis technique, using the GO database as biological knowledge.

For each pipeline, we obtained a list of discriminant GO terms as well as a list of selected genes that were compared with the benchmark lists, defined following the workflow of Fig. 3 (see Materials and Methods) and that represents acquired knowledge about PD. Based on the comparison between the benchmark lists and the lists identified by both pipelines, we calculated the precision, recall and F-measures (Tables 1, 2). In the GO terms case, it is evident that the recall of KDVS was higher with respect to the Standard pipeline, considering a comparable precision. In the gene case, both the precision and the recall values were higher for KDVS than for the other pipeline. The meaning of this high recall for KDVS, considering either GO terms and genes, is that this approach identified most of the relevant GO terms or genes in comparison with the Standard approach. The F-measures confirmed the better performance of KDVS in comparison with the other approach.

Finally, from the three–way comparison, one new gene and thirteen GO terms emerged in the spotlight. The gene is ATXN1, known to be associated with ADCA, another neurodegenerative disease. The terms are a combination of processes or functions already known to be involved in PD and not yet related with the disease.

These pinpointed entities were derived from the application of two different pipelines to the same data set related to early onset PD, and emerged from the comparison with the verified knowledge on the disease (*benchmark lists*). This suggests that a direct exploration of the possible impact of these entities in the development of PD should be performed.

# References

1. Avramopoulos, D.: Genetics of alzheimer's disease: recent advances. Genome Med. **1**(3), 34 (2009)
2. Mrzljak, L., Munoz-Sanjuan, I.: Therapeutic strategies for huntingtons disease. Current Topics in Behavioral Neurosciences, pp. 1–41. Springer, Heidelberg (2013)
3. Zycinski, G., Barla, A., Verri, A.: Svs: Data and knowledge integration in computational biology. In: Conference Proceedings of the IEEE Engineering in Medicine and Biology Society 2011, pp. 6474–6478 (2012)

4. Zycinski, G., Barla, A., Squillario, M., Sanavia, T., Di Camillo, B., Verri, A.: Knowledge driven variable selection (kdvs) - a new approach to enrichment analysis of gene signatures obtained from high-throughput data. Source Code Biol. Med. **8**(1), 2 (2013)
5. Guyon, I., Elisseeff, A.: An introduction to variable and feature selection. J. Mach. Learn. Res. **3**, 1157–1182 (2003)
6. Da Huang, W., Sherman, B.T., Lempicki, R.A.: Systematic and integrative analysis of large gene lists using david bioinformatics resources. Nat. Protoc. **4**(1), 44–57 (2009)
7. Zhang, B., Kirov, S., Snoddy, J.: Webgestalt: an integrated system for exploring gene sets in various biological contexts. Nucleic Acids Res. **33**, W741–W748 (2005)
8. Ashburner, M., Ball, C.A., Blake, J.A., Botstein, D., Butler, H., Cherry, J.M., Davis, A.P., Dolinski, K., Dwight, S.S., Eppig, J.T., Harris, M.A., Hill, D.P., Issel-Tarver, L., Kasarskis, A., Lewis, S., Matese, J.C., Richardson, J.E., Ringwald, M., Rubin, G.M., Sherlock, G.: Gene ontology: tool for the unification of biology. the gene ontology consortium. Nat. Genet. **25**(1), 25–29 (2000)
9. Kanehisa, M., Goto, S.: Kegg: kyoto encyclopedia of genes and genomes. Nucleic Acids Res. **28**(1), 27–30 (2000)
10. Barla, A., Mosci, S., Rosasco, L., Verri, A.: A method for robust variable selection with signicance assessment. In: Proceedings of ESANN 2008 (2008)
11. Squillario, M., Barla, A.: A computational procedure for functional characterization of potential marker genes from molecular data: Alzheimer's as a case study. BMC Med. Genomics **4**, 55 (2011)
12. Fardin, P., Barla, A., Mosci, S., Rosasco, L., Verri, A., Varesio, L.: The l1–l2 regularization framework unmasks the hypoxia signature hidden in the transcriptome of a set of heterogeneous neuroblastoma cell lines. BMC Genom. **10**, 474 (2009)
13. Edgar, R., Domrachev, M., Lash, A.E.: Gene expression omnibus: Ncbi gene expression and hybridization array data repository. Nucleic Acids Res. **30**(1), 207–210 (2002)
14. Scherzer, C.R., Eklund, A.C., Morse, L.J., Liao, Z., Locascio, J.J., Fefer, D., Schwarzschild, M.A., Schlossmacher, M.G., Hauser, M.A., Vance, J.M., Sudarsky, L.R., Standaert, D.G., Growdon, J.H., Jensen, R.V., Gullans, S.R.: Molecular markers of early parkinson's disease based on gene expression in blood. Proc. Natl. Acad. Sci. U. S. A. **104**(3), 955–960 (2007)
15. Irizarry, R.A., Bolstad, B.M., Collin, F., Cope, L.M., Hobbs, B., Speed, T.P.: Summaries of affymetrix genechip probe level data. Nucleic Acids Res. **31**(4), e15 (2003)
16. De Mol, C., De Vito, E., Rosasco, L.: Elastic net regularization in learning theory. J. Complex. **25**, 201–230 (2009)
17. Zou, H., Hastie, T.: Regularization and variable selection via the elastic net. J. Roy. Stat. Soc. Ser. B **67**, 301–320 (2005)
18. Ambroise, C., McLachan, G.: Selection bias in gene extraction on the basis of microarray gene-expression data. Proc. Natl. Acad. Sci. U. S. A. **99**(10), 6562–6566 (2002)
19. Gentleman, R.C., Carey, V.J., Bates, D.M., Bolstad, B., Dettling, M., Dudoit, S., Ellis, B., Gautier, L., Ge, Y., Gentry, J., Hornik, K., Hothorn, T., Huber, W., Iacus, S., Irizarry, R., Leisch, F., Li, C., Maechler, M., Rossini, A.J., Sawitzki, G., Smith, C., Smyth, G., Tierney, L., Yang, J.Y., Zhang, J.: Bioconductor: open software development for computational biology and bioinformatics. Genome Biol. **5**(10), R80 (2004)
20. Leung, Y.F., Cavalieri, D.: Fundamentals of cdna microarray data analysis. Trends Genet. **19**(11), 649–659 (2003)

21. Subramanian, A., Tamayo, P., Mootha, V.K., Mukherjee, S., Ebert, B.L., Gillette, M.A., Paulovich, A., Pomeroy, S.L., Golub, T.R., Lander, E.S., Mesirov, J.P.: Gene set enrichment analysis: a knowledge-based approach for interpreting genome-wide expression profiles. PNAS **102**(43), 15545–15550 (2005)
22. Yu, W., Wulf, A., Liu, T., Khoury, M.J., Gwinn, M.: Gene prospector: an evidence gateway for evaluating potential susceptibility genes and interacting risk factors for human diseases. BMC Bioinf. **9**, 528 (2008)
23. Jomova, D.V.K., Lawson, M., Valko, M.: Metals, oxidative stress and neurodegenerative disorders. Mol Cell Biochem. **345**, 91–104 (2010)
24. Gundersen, V.: Protein aggregation in parkinson's disease. Acta. Neurol. Scand. Suppl. **122**, 82–87 (2010)
25. Ronchi, C., Cartelli, D., Rodighiero, S., Maggioni, M.G., Cappelletti, G., Giavini, E.: J. Neurochemestry **115**(1), 247–258 (2010)
26. Rohn, T.T., Catlin, L.W.: Immunolocalization of influenza a virus and markers of inflammation in the human parkinson's disease brain. PLoS One **6**(5), e20495 (2011)
27. Clayton, D.F., George, J.M.: The synucleins: a family of proteins involved in synaptic function, plasticity, neurodegeneration and disease. Trends Neurosci. **21**(6), 249–254 (1998)
28. Tezenas du Montcel, S., Charles, P., Goizet, C., Marelli, C., Ribai, P., Vincitorio, C., Anheim, M., Guyant-Marchal, L., Le Bayon, A., Vandenberghe, N., Tchikviladz, M., Devos, D., Le Ber, I., N'Guyen, K., Cazeneuve, C., Tallaksen, C., Brice, A., Durr, A.: Factors influencing disease progression in autosomal dominant cerebellar ataxia and spastic paraplegia. Arch. Neurol. **69**(4), 500–508 (2012)

# Author Index

Printed in the United States
By Bookmasters